DECODING ARTICLE 6 OF THE PARIS AGREEMENT

VERSION II

DECEMBER 2020

ASIAN DEVELOPMENT BANK

ADB

© 2020 Asian Development Bank
6 ADB Avenue, Mandaluyong City, 1550 Metro Manila, Philippines
Tel +63 2 8632 4444; Fax +63 2 8636 2444
www.adb.org

Some rights reserved. Published in 2020.

ISBN 978-92-9262-619-8 (print); 978-92-9262-620-4 (electronic); 978-92-9262-621-1 (ebook)
Publication Stock No. TCS200411-2
DOI: http://dx.doi.org/10.22617/TCS200411-2

The views expressed in this publication are those of the authors and do not necessarily reflect the views and policies of the Asian Development Bank (ADB) or its Board of Governors or the governments they represent.

ADB does not guarantee the accuracy of the data included in this publication and accepts no responsibility for any consequence of their use. The mention of specific companies or products of manufacturers does not imply that they are endorsed or recommended by ADB in preference to others of a similar nature that are not mentioned.

By making any designation of or reference to a particular territory or geographic area, or by using the term "country" in this document, ADB does not intend to make any judgments as to the legal or other status of any territory or area.

Please contact pubsmarketing@adb.org if you have questions or comments with respect to content, or if you wish to obtain copyright permission for your intended use that does not fall within these terms, or for permission to use the ADB logo.

Corrigenda to ADB publications may be found at http://www.adb.org/publications/corrigenda.

Notes:
In this publication, "$" refers to United States dollars.
ADB recognizes "Korea" as the Republic of Korea.

Contents

Figures and Boxes

FIGURES

BOXES

Foreword

The pandemic caused by the coronavirus disease (COVID-19) has disrupted billions of lives and livelihoods across the world in 2020. The Asian Development Bank (ADB) estimates the global economic impact of the COVID-19 pandemic to reach $4.8–$7.4 trillion, equivalent to 5.5%–8.7% of the world's cumulative gross domestic product (GDP) in 2020, and an additional $3.1–$5.4 trillion in 2021 (or 3.6%–6.3% of GDP). The Asia and Pacific region* is expected to account for 28% of the overall decline in the global output due to the pandemic, with a loss of 6.0%–9.5% of regional GDP in 2020 and 3.6%–6.3% in 2021.

As governments adopt response and recovery measures to mitigate the impacts of the pandemic, it is important that long-term solutions are taken into consideration. Fortunately, committing to long-term measures is nothing new to many countries, as 189 out of 197 Parties to the United Nations Framework Convention on Climate Change (UNFCCC) are also Parties to the Paris Agreement—a landmark framework that sets out long-term goals to strengthen the global response to address climate change and limit global warming by the end of the 21st century.

There is now an opportunity to align COVID-19 recovery plans to countries' climate actions as there is an increasing recognition at the political and grassroots levels for the recovery to "build back better." It is through this momentum on "green recovery" that there is hope in putting the world on a more sustainable path toward economic recovery from the pandemic.

Addressing climate change is a key part of the more sustainable path. The Asia and Pacific region plays a key role in implementing climate actions to limit global warming to well below 2°C. In 2018, the region generated about 46% of global carbon dioxide (CO_2) emissions. This is worrying since the overall concentration of CO_2 in the atmosphere is still rising and the collective ambition expressed by the sum of nationally determined contributions (NDCs) is largely insufficient to meet the Paris Agreement goals. Even though there has been a momentary dip in greenhouse gas emissions during the lockdowns instituted to control the increase of COVID-19 cases, emissions will bounce back to pre-pandemic levels once countries start to ease restrictions, if no green recovery measures are put in place.

---

* Asia and Pacific region refers to the 46 members of the Asian Development Bank listed below. Central Asia comprises Armenia, Azerbaijan, Georgia, Kazakhstan, Kyrgyz Republic, Tajikistan, Turkmenistan, and Uzbekistan. East Asia comprises Hong Kong, China; Mongolia; People's Republic of China; Republic of Korea; and Taipei,China. South Asia comprises Afghanistan, Bangladesh, Bhutan, India, Maldives, Nepal, Pakistan, and Sri Lanka. Southeast Asia comprises Brunei Darussalam, Cambodia, Indonesia, Lao People's Democratic Republic, Malaysia, Myanmar, Philippines, Singapore, Thailand, Timor-Leste, and Viet Nam. The Pacific comprises Cook Islands, Federated States of Micronesia, Fiji, Kiribati, Marshall Islands, Nauru, Niue, Palau, Papua New Guinea, Samoa, Solomon Islands, Tonga, Tuvalu, and Vanuatu.

In this context, the Paris Agreement, through its Article 6, recognizes the role of international cooperation in increasing climate ambition. The Paris Agreement has also reignited the hope for a more coherent international carbon market—and one that assists Parties in increasing their ambition in relation to their NDCs. Article 6 and the introduction of cooperative approaches are the foundations for such rebuilding of the international carbon market architecture. In a study conducted by the International Emissions Trading Association in 2019, it was found that countries can reduce costs in implementing their stated NDC commitments if they cooperate under the principles of Article 6 of the Paris Agreement. A total of $250 billion per year in 2030 can be saved globally from improved economic efficiency— savings which could be diverted by country governments to other priorities, or channeled to enhance climate mitigation actions which will result in an increase in global carbon emissions mitigation by up to 5 gigatons of CO_2 equivalent per year.

While countries can start to cooperate on carbon market instruments today, the development of bilateral and multilateral approaches and mechanisms will benefit from a clear international framework. Unfortunately, this is not yet in place. Hampered by the sheer complexity of the issues, and recently also delayed by the pandemic, international negotiations on Article 6 will not reach a resolution until the 26th session of the Conference of the Parties to the UNFCCC (COP26) in November 2021.

This delay is both a blessing and a curse. On the one hand, it means that stakeholders that want to engage in the new carbon market will not have regulatory certainty regarding some of the key issues. This is above all a challenge for the new mechanism under Article 6.4, which will not be operational until there is a decision on Article 6 including the establishment of a supervisory body. On the other hand, the delay has given countries and stakeholders time to digest and discuss technical and political issues of Article 6 before the resumption of the negotiations, and before countries begin detailed preparations for participating in Article 6 approaches.

It is in this light that ADB is now publishing *Decoding Article 6 of the Paris Agreement Version II*, a sequel to the first version which ADB published in 2018 leading up to COP24. Version II shows that there has been progress in the negotiations, and that there is a better understanding among countries of the options and alternatives to engaging in cooperative approaches under Article 6. Hopefully, the knowledge gained from this publication can help Article 6 negotiators, which can then support the negotiations to lead to a decision for the Article 6 rulebook in late 2021. We hope that countries will take the opportunity to make use of the cooperative approaches in Article 6, so as to enhance the green recovery from the pandemic and set the countries on a low-carbon development pathway.

Woochong Um
Director General
Sustainable Development and Climate Change Department
Asian Development Bank

Preface

More than 2 decades ago, the public debate focused largely on how to design a single global market for trading carbon units as the key instrument for addressing global climate change. The argument emphasized that since 1 ton of a greenhouse gas emitted anywhere in the world has the same climate change consequences for everyone, a single global market would be an economically desirable outcome, equalizing incentives to reduce emissions everywhere. Today, this late-1990s dream of a top-down global design seems far away, if not impossible. Instead, we see a multiplicity of regional, national, and even subnational markets emerging.

This trend of multiplicity and fragmentation is reflected in the bottom-up architecture of the Paris Agreement. The decentralized approach under Article 6.2 offers countries flexibility and a choice in their approach. While such an approach opens for innovation and national adaptation of carbon market instruments, the lack of a centralized approach and harmonized standards for monitoring, reporting, and verification means that comparability could be difficult to attain.

The emerging complexity is also reflected in the ongoing Article 6 negotiations. The continuing struggle to come to an agreement over guidance and rules for Article 6 has created uncertainty over when the rules for international carbon markets under the Paris Agreement will be set, and what the implications for participants will be.

At the same time, since many elements in the Paris Agreement are new, it should not be surprising that the process will take time. The Kyoto Protocol was established in 1997, yet it took 4 years to agree on the detailed rules for carbon market mechanisms, and these mechanisms were further developed and tested well in advance of the start of first commitment period of the Kyoto Protocol in 2008. In this perspective, seeing Article 6 mechanisms in operation within 10 years from the adoption of the Paris Agreement is perhaps realistic.

Article 6 of the Paris Agreement provides for developing and using mechanisms designed by countries or using the Article 6.4 mechanism that is subject to centralized oversight by the United Nations Framework Convention on Climate Change (UNFCCC). The latter is likely to be attractive for countries with limited capacity or will to develop and design mechanisms on their own, or those that prefer to have an UN-quality stamp on the mitigation outcomes achieved.

Article 6.2 reflects the bottom-up ethos of the Paris Agreement making it possible to use a variety of designs for carbon market instruments. This enables innovation in how carbon market mechanisms are implemented, as well as the purposes for which they are used. This means that domestic carbon pricing or offsetting schemes can be used for export of carbon credits as long as they comply with the accounting guidance under Article 6.2.

Agreeing on the rules and guidance for these two approaches is not the only challenge for negotiators. In the current negotiations, some issues are old, others have emerged more recently, or, depending on who is asked, have reemerged. One issue that has emerged as contentious is the transfer of different elements of the Clean Development Mechanism (CDM) from the Kyoto Protocol framework to the new Paris Agreement. Another more recent controversy relates to how the respective approach should contribute to adaptation, which for the CDM was managed by a Share of Proceeds. Still at the centerpiece of discussion is the relationship between Article 6.2 and Article 6.4, and how robust accounting should be implemented.

The Asian Development Bank (ADB) provides capacity-building and policy development support to its developing member countries (DMCs) through the Article 6 Support Facility under its Carbon Market Program. One of the key objectives of this effort is to help DMCs navigate the ongoing negotiations. *Decoding Article 6 of the Paris Agreement* has been an important document in this process, taking a comprehensive view of negotiation issues and outlining different views and interpretations.

Decoding Article 6 of the Paris Agreement Version II could be equally important. The momentum in discussions needs to be maintained, while it also is important to take stock of developments that have occurred since 2018. As with the first version, it is our hope that this publication will be useful to build an in-depth understanding of Article 6 and engage stakeholders in fruitful discussions to further shape climate actions. ADB hopes that this will enable DMCs to contribute to the development of the new rules and eventually lead them to take advantage of the new carbon markets under the Paris Agreement.

Preety Bhandari
Chief of Climate Change and Disaster
Risk Management Thematic Group and
Director, Climate Change and Disaster Risk
Management Division
Sustainable Development and Climate
Change Department
Asian Development Bank

Virender Kumar Duggal
Principal Climate Change Specialist
Fund Manager-Future Carbon Fund
Sustainable Development and Climate
Change Department
Asian Development Bank

Acknowledgments

This knowledge product has been developed by the Article 6 Support Facility under the Carbon Market Program of the Asian Development Bank (ADB) within its Sustainable Development and Climate Change Department.

Virender Kumar Duggal, principal climate change specialist, Climate Change and Disaster Risk Management Division, ADB, conceptualized and guided development of this knowledge product.

The knowledge product has been developed with the technical research and analysis conducted by Andrei Marcu, whose contribution is greatly appreciated.

The design and development of this knowledge product has greatly benefited from advice from various technical experts under ADB's Carbon Market Program–Naresh Badhwar, Pedro Martins Barata, Rastraraj Bhandari, Deborah Cornland, Hannah Ebro, Brij Mohan, and Johan Nylander, all of which is deeply appreciated. The timely publication of this paper was made possible by the valuable advice and support from Janet Arlene Amponin, Esmyra Javier, and Ghia Villareal.

This knowledge product has hugely benefited from the peer review conducted by Seoyoung Lim of the Korea Environment Corporation (K-eco), which is sincerely acknowledged and appreciated.

This publication was made possible with the excellent coordination and administrative support from Anna Liza Cinco, Ken Edward Concepcion, Kristine Lim Ang, and Jeanette Morales.

The preparation of this publication has immensely benefited from the diligent support from Layla Amar (editing), Edith Creus (cover design), Joseph Manglicmot (typesetting/layout), Lawrence Casiraya (proofreading), and Jess Alfonso Macasaet (page proof checking), all of which is sincerely appreciated.

The support provided by the publishing team in ADB's Department of Communications and the Printing Services Unit in ADB's Office of Administrative Services in publishing this knowledge product is also acknowledged.

Abbreviations

A6.4U	–	Article 6.4 Emission Reduction Units
A6.4M	–	Article 6.4 Mechanism
BAU	–	business-as-usual
BTR	–	biennial transparency report
BUR	–	Biennial Update Report
CARP	–	Centralized Accounting and Reporting Platform
CDM	–	Clean Development Mechanism
CER	–	certified emissions reductions
CMA	–	Conference of the Parties serving as the Meeting of the Parties to the Paris Agreement
CO_2e	–	carbon dioxide equivalent
COP	–	Conference of the Parties
ETF	–	enhanced transparency framework
ETS	–	emissions trading system
GHG	–	greenhouse gas
ICAO	–	International Civil Aviation Organization
IPCC	–	Intergovernmental Panel on Climate Change
ITMO	–	internationally transfered mitigation outcome
LDC	–	least-developed country
MPG	–	modalities, procedures, and guidelines

NDC   –   nationally determined contribution

NMA   –   nonmarket approach

OMGE   –   overall mitigation of global emissions

SBSTA   –   Subsidiary Body for Scientific and Technological Advice

SIDS   –   small island developing state

SOP   –   share of proceeds

tCO_2e –   ton of carbon dioxide equivalent

TER   –   technical expert review

UNFCCC   –   United Nations Framework Convention on Climate Change

1. Introduction

1.1 Overview of the Paris Agreement

The Paris Agreement[1] was adopted at the 21st session of the Conference of the Parties (COP21) in 2015 in Paris. It represents a new and important step in the evolution of the climate change policy agreements which includes the United Nations Framework Convention on Climate Change (UNFCCC) and the Kyoto Protocol as other important landmarks.

The ultimate objective of the UNFCCC is to achieve stabilization of greenhouse gas (GHG) concentrations in the atmosphere at a level that would prevent dangerous anthropogenic interference with the climate system. In the UNFCCC, all Parties will cooperate in this objective. Developed countries are to adopt national policies, take measures, and provide new and additional financial resources.

The 1997 Kyoto Protocol[2] spells out a concrete way forward. The Kyoto Protocol includes legally binding targets or commitments to reduce or limit GHG emissions and more stringent reporting and review requirements for developed countries (called Annex 1 countries). There were no obligations on developing countries. It is important to note that in the case of the Kyoto Protocol, all obligations by Annex 1 countries were expressed in the same way, through a budget with a reference level 1990 emissions by that Party.

For the first commitment period of the Kyoto Protocol, the objective was –5% (in 2008–2012) to 1990 level; in reality, about –20% was achieved. The objective for the second commitment period was –18%, for the period 2013–2020. However, the conditions for the entry into force of the second commitment period were met in the Fall of 2020. The Kyoto Protocol also included provisions for flexibility mechanisms through emissions trading, the Clean Development Mechanism (CDM), and joint implementation.

The Paris Agreement is an agreement whereby all Parties take on some commitment expressed through their nationally determined contribution (NDC). There is no compliance and enforcement, but the Paris Agreement incorporates a significant amount of transparency, with flexibility for developing countries. It introduces a number of new

---

[1]   UNFCCC. 2015. Paris Agreement. https://unfccc.int/files/essential_background/convention/application/pdf/english_paris_agreement.pdf.

[2]   UNFCCC. 1997. Kyoto Protocol to the United Nations Framework Convention on Climate Change. 10 December. https://unfccc.int/sites/default/files/resource/docs/cop3/l07a01.pdf.

concepts and builds on the experience of the UNFCCC and the Kyoto Protocol, by providing the flexibility of international cooperation, including market approaches.

The Paris Agreement has global objectives and Party contributions to the global objectives expressed through the NDCs. It aims to strengthen the global response to the threat of climate change by keeping a global temperature rise this century well below 2°C above pre-industrial levels and to pursue efforts to limit the increase even further to 1.5°C.

The Paris Agreement also provides a common framework for Parties to take ambitious efforts to combat climate change and adapt to its effects, with enhanced support to developing countries and requiring all Parties to put forward their best efforts through NDCs and to strengthen these efforts over time. The commitment included in the NDC as well as the way it is expressed is "nationally determined."

The goal of the Paris Agreement is to strengthen the ability of countries to deal with the impacts of climate change. To reach these ambitious goals, appropriate financial flows, a new technology framework, and an enhanced capacity-building framework will be put in place, supporting action by developing countries and the most vulnerable countries, in line with their own national objectives. The Paris Agreement provides for enhanced transparency of action and support through a more robust transparency framework. It also introduces a global stocktake every 5 years to assess the collective progress toward achieving the purpose of the agreement and to inform further actions by the Parties.

The Paris Agreement incorporates the following articles:

(i)   Article 2 defines the objective,
(ii)   Articles 3 and 4 define the nationally determined contributions,
(iii)   Article 5 addresses sinks and reservoirs of GHGs,
(iv)   Article 6 is about voluntary international cooperation,
(v)   Article 7 addresses adaptation to climate change,
(vi)   Article 8 focuses on the issues of loss and damage,
(vii)   Article 9 is on financial resources,
(viii) Article 10 deals with technology development and transfer,
(ix)   Article 11 and 12 discuss capacity building,
(x)   Article 13 establishes a new transparency framework,
(xi)   Article 14 establishes the new global stocktake, and
(xii)   Article 15 establishes a compliance mechanism.

1.2 Scope of Article 6

Box 1: Article 6 of the Paris Agreement

1. Parties recognize that some Parties choose to pursue voluntary cooperation in the implementation of their nationally determined contributions to allow for higher ambition in their mitigation and adaptation actions, and to promote sustainable development and environmental integrity.

2. Parties shall, where engaging on a voluntary basis in cooperative approaches that involve the use of internationally transferred mitigation outcomes toward nationally determined contributions, promote sustainable development and ensure environmental integrity and transparency, including in governance, and shall apply robust accounting to ensure, inter alia, the avoidance of double counting, consistent with guidance adopted by the Conference of the Parties serving as the meeting of the Parties to this Agreement.

3. The use of internationally transferred mitigation outcomes to achieve nationally determined contributions under this Agreement shall be voluntary and authorized by participating Parties.

4. A mechanism to contribute to the mitigation of greenhouse gas emissions and support sustainable development is hereby established under the authority and guidance of the Conference of the Parties serving as the meeting of the Parties to this Agreement for use by Parties on a voluntary basis. It shall be supervised by a body designated by the Conference of the Parties serving as the meeting of the Parties to this Agreement, and shall aim:

   (a) to promote the mitigation of greenhouse gas emissions while fostering sustainable development;
   (b) to incentivize and facilitate participation in the mitigation of greenhouse gas emissions by public and private entities authorized by a Party;
   (c) to contribute to the reduction of emission levels in the host Party, which will benefit from mitigation activities resulting in emission reductions that can also be used by another Party to fulfil its nationally determined contribution; and
   (d) to deliver an overall mitigation in global emissions.

5. Emission reductions resulting from the mechanism referred to in paragraph 4 of this Article shall not be used to demonstrate achievement of the host Party's nationally determined contribution if used by another Party to demonstrate achievement of its nationally determined contribution.

6. The Conference of the Parties serving as the meeting of the Parties to this Agreement shall ensure that a share of the proceeds from activities under the mechanism referred to in paragraph 4 of this Article is used to cover administrative expenses as well as to assist developing country Parties that are particularly vulnerable to the adverse effects of climate change to meet the costs of adaptation.

7. The Conference of the Parties serving as the meeting of the Parties to this Agreement shall adopt rules, modalities and procedures for the mechanism referred to in paragraph 4 of this Article at its first session.

continued on next page

Box 1 continued

8. Parties recognize the importance of integrated, holistic and balanced nonmarket approaches (NMAs) being available to Parties to assist in the implementation of their nationally determined contributions, in the context of sustainable development and poverty eradication, in a coordinated and effective manner, including through, inter alia, mitigation, adaptation, finance, technology transfer and capacity-building,as appropriate. These approaches shall aim to:

(a) promote mitigation and adaptation ambition,
(b) enhance public and private sector participation in the implementation of nationally determined contributions, and
(c) enable opportunities for coordination across instruments and relevant institutional arrangements.

9. A framework for NMAs to sustainable development is hereby defined to promote the NMAs referred to in paragraph 8 of this Article.

Source: UNFCCC. The Paris Agreement. https://unfccc.int/files/essential_background/convention/application/pdf/english_paris_agreement.pdf.

Article 6 of the Paris Agreement (see Box 1) addresses international voluntary cooperation and was, to a large degree, a surprise in Paris at COP21. It is often called the "market article" while in reality it is very much about international cooperation, and includes the special case where the cooperation includes the transfer of mitigation outcomes that will be used for meeting the NDC in a Party other than where the mitigation outcome took place. The fact that it addresses the treatment of mitigation outcomes transferred internationally allows for the creation of carbon markets by those Parties that wish to avail themselves of that opportunity by setting up such markets. It must be noted that there is no direct reference to markets in the Article 6 text.

Article 6 is seen by most negotiators and stakeholders as comprised of four modules or components:[3]

(i)   Paragraph 6.1.

This paragraph is about the general concept that Parties may cooperate, on a voluntary basis, in the implementation of their NDCs. Article 6 is designed to cover existing types of cooperation, and those that are yet to emerge. This cooperation does not need approval by a body under the Paris Agreement. Rather, it is to be noted, acknowledged, and recognized. This is important as it reinforces the decentralized and bottom–up nature and ethos of the Paris Agreement governance. Other important language in this paragraph is that around "ambition." The reference to "allow for higher ambition" is also important in the formulation of this paragraph. Some Paris Agreement drafts on these issues (cooperation, transfers, markets, etc.) referred to the need to "enhance mitigation ambition."

---

[3]   ADB. 2018. *Decoding Article 6 of the Paris Agreement*. Manila.

The use of the word "enhance" was disputed by many Parties that wanted to make sure that cooperative approaches could be used to achieve what was in their intended NDCs at the time. They felt that the use of the word "enhance" could be interpreted as the need to increase the level of ambition in their current NDC before they could make use of cooperative approaches. "Allow" has a "facilitative" connotation, while "enhance" would seem to require an active act of increasing the level of ambition.

(ii)   Transfers of mitigation outcomes (paragraphs 6.2–6.3).

Paragraphs 6.2–6.3 cover the concept that when Parties engage in Cooperative Approaches that involve mitigation outcomes being transferred internationally and used toward the NDC of another Party than where the mitigation outcome was produced, they need to respect the guidance on accounting and avoidance of double counting decided by the COP serving as the meeting of the Parties to the Paris Agreement (CMA). Further, any international transfer of mitigation outcomes will also need to respect two other requirements: "Parties shall [...] promote sustainable development and ensure environmental integrity and transparency, including in governance."

These paragraphs are not about markets *per se*, but they create a framework on how to account for transfers between Parties and what conditions need to be met. Important is that these internationally transferred mitigation outcomes (ITMOs) can be a result of any mitigation approaches (e.g., mechanism, procedure, or protocol). Thus, there is no requirement that these approaches operate under the authority of the COP. Essentially it is whatever the Parties involved will agree.

These paragraphs do not impose any limitation on as to what constitutes an ITMO. This broad scope is supported by the "institutional memory" of the Paris Agreement negotiations. Should limitations be introduced, they will essentially be an additional "boundary" or limitation which Parties to the Paris Agreement agree in the operationalization of Article 6, but currently have no "hook" in the current text.

(iii)   Mechanism to contribute to mitigation and support sustainable development (paragraphs 6.4–6.7).

Paragraphs 6.4-6.7 establish a mechanism to produce mitigation outcomes and support sustainable development. It operates under the authority of the COP. This mechanism will produce mitigation outcomes that can then be used to achieve the NDC target of another Party. An issue still under debate is whether the scope of these paragraphs is limited to a CDM-like mechanism, or if it is much broader in scope.

Opening for a broader scope seems to receive support from the historical evolution of the text, from the submissions on Article 6 to the Subsidiary Body for Scientific and Technological Advice (SBSTA), as well as from positions expressed in formal and informal discussions.

(iv)   Framework for nonmarket approaches (paragraphs 6.8 and 6.9).

The establishment of a framework for NMAs will aim to result in a variety of NMAs being implemented, a form of governance of the framework, and a work program of the framework. What to be covered under this part of Article 6 is still largely unknown, but some focus is starting to emerge. One area seems to be coordination of different nonmarket cooperation approaches.

Alternative ideas that have surfaced suggest that Articles 6.8 and 6.9 should be complementary to other provisions in the Paris Agreement, including those in Articles 6.2 to 6.7. The aim would be to ensure the sustainability of mitigation approaches, as well as to address issues of global competitiveness in a cooperative manner, which relates to Article 4.15 of the Paris Agreement.

It is important to recall that when Parties were negotiating the Paris Agreement, they wanted to provide alternatives that they could use in cooperating internationally in implementing the Paris Agreement and their respective NDCs.

The two governance modes, one more centralized and another less centralized, were provided in a very deliberate way, providing options for participating in markets, allowing Parties to have choices.

As part of the operationalization of the Paris Agreement, Parties will negotiate the details of all these paragraphs and will agree on the level of governance centralization for Articles 6.2 and 6.4. However, this deliberate decision to allow Parties a choice between a procedure with more centralized UNFCCC-centric governance, or a more bottom-up approach with more responsibility on the cooperating Parties, is something that needs to be always recalled.

1.3   Draft Negotiating Text

As background to this paper, it is important to mention that three versions of a draft text for matters relating to Article 6 of the Paris Agreement were put forward by the president of COP25 in Madrid. Throughout this paper, quotations are taken from version 3 of the draft text.[4]

The draft text has two components: the Decision and the Annex. The Decision describes actions to implement Article 6 and other future actions by the COP related to Article 6. The Annex contains the actual text of the rulebook.

---

[4]   Draft decision texts on guidance on cooperative approaches referred to in Article 6, paragraph 2, of the Paris Agreement, available at https://unfccc.int/documents/204687 (third iteration, 15 December); rules, modalities and procedures for the mechanism established by Article 6, paragraph 4, of the Paris Agreement, available at https://unfccc.int/documents/204686 (third iteration, 15 December); and the work program under the framework for NMAs referred to in Article 6, paragraph 8, of the Paris Agreement, available at https://unfccc.int/documents/204667 (third iteration, 15 December).

1.4 A Brief History of Negotiations Since the 21st Session of the Conference of the Parties

When the Paris Agreement was presented to the plenary, the inclusion of significant provisions related to markets under Article 6 came as a surprise to most stakeholders—the rate of progress on carbon market issues had stalled for many negotiating sessions and even in Paris, stakeholders did not hold out much hope. The Paris outcome left many in a very upbeat mood and with a hope for fast progress.

The Paris Agreement was accompanied by a COP decision (1/CP.21)[5] which elaborated a series of work programs to operationalize different articles in the Paris Agreement. Three work programs addressed specific provisions under Article 6:

(i)   Para. 36 is related to the guidance on accounting and the avoidance of double counting included in Article 6.2:

[The Conference of the Parties] requests the Subsidiary Body for Scientific and Technological Advice to develop and recommend the guidance referred to under Article 6, paragraph 2, of the Agreement for consideration and adoption by the Conference of the Parties serving as the meeting of the Parties to the Paris Agreement at its first session, including guidance to ensure that double counting is avoided on the basis of a corresponding adjustment by Parties for both anthropogenic emissions by sources and removals by sinks covered by their nationally determined contributions under the Agreement.

(ii)   Para. 37 is related to the development of modalities and procedures for the new mechanism outlined under Article 6.4:

[The Conference of the Parties] recommends that the Conference of the Parties serving as the meeting of the Parties to the Paris Agreement adopt rules, modalities and procedures for the mechanism established by Article 6, paragraph 4, of the Agreement on the basis of:

(a)   *Voluntary participation authorized by each Party involved;*
(b)   *Real, measurable, and long-term benefits related to the mitigation of climate change;*
(c)   *Specific scopes of activities;*
(d)   *Reductions in emissions that are additional to any that would otherwise occur;*
(e)   *Verification and certification of emission reductions resulting from mitigation activities by designated operational entities; and*
(f)   *Experience gained with and lessons learned from existing mechanisms and approaches adopted under the Convention and its related legal instruments.*

---

[5]   UNFCCC. 2016. FCCC/CP/2015/10/Add.1. Available at: https://unfccc.int/sites/default/files/resource/docs/2015/cop21/eng/10a01.pdf.

Para.39 requests SBSTA to undertake a work program related to NMAs:

" with the objective of considering how to enhance linkages and create synergy between, inter alia, mitigation, adaptation, finance, technology transfer and capacity-building, and how to facilitate the implementation and coordination of nonmarket approaches."

Paras.38 and 40 mandate the SBSTA to present its recommendations on the decisions and annexes regarding both the rules on Article 6.4 (para.38) and the work program under Article 6.8 (para.40) to the first CMA. These were then not adopted at CMA1 (COP22) nor at CMA2 (COP23).

These work programs were meant to be adopted in Katowice at the 24th session of the COP (COP24) as part of the Paris Agreement rulebook. The outcome of COP24 was a great disappointment to those who worked on Article 6 as this was the only part of the Paris Agreement that was not adopted. Figure 1 below depicts an overview of key negotiation events and events/deadlines scheduled under the Paris Agreement.

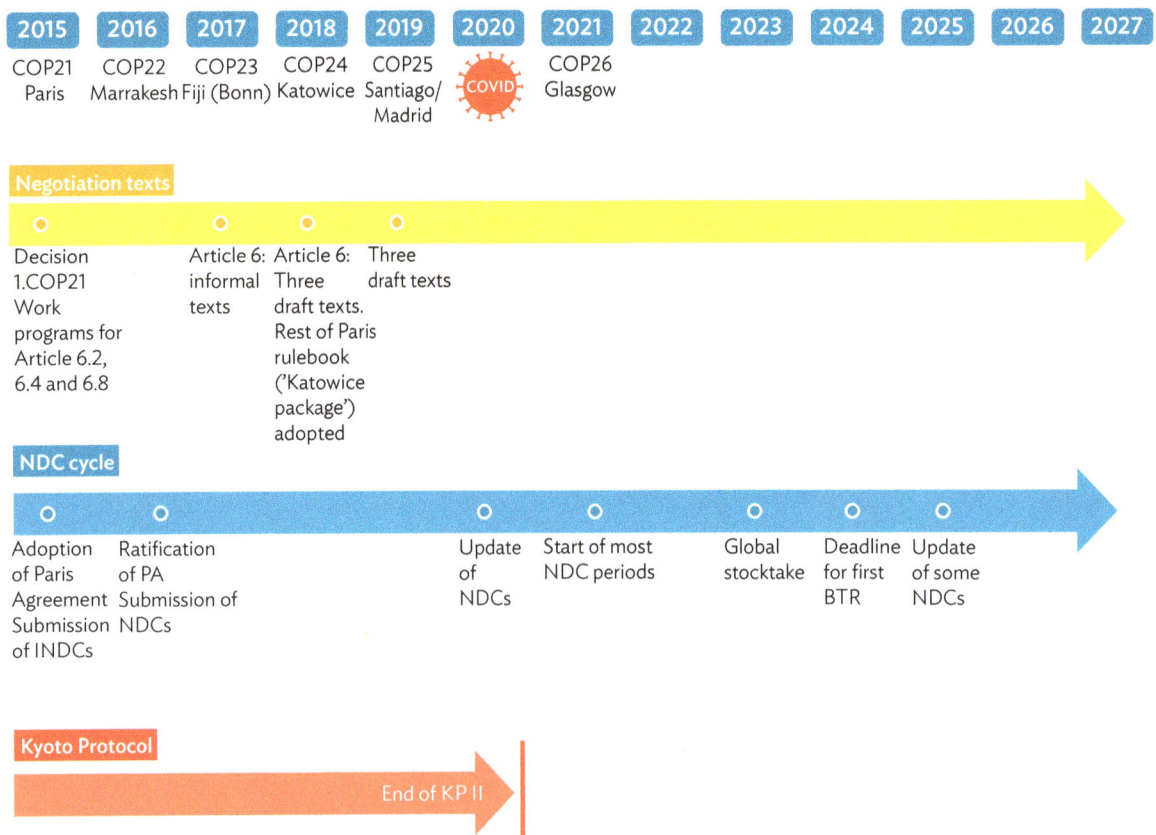

Figure 1: Overview of Negotiations and Key UNFCCC Events

BTR = biennial transparency report, COP = Conference of the Parties, INDC = intended nationally determined contribution, KP = Kyoto Protocol, NDC = nationally determined contribution, PA = Paris Agreement, UNFCCC = United Nations Framework Convention on Climate Change.

Source: ADB

1.4.1 Outcomes of Article 6 from the 24th Session of the Conference of the Parties

The final outcome of Article 6 in the Katowice rulebook was largely procedural. It was part of the Katowice Climate Package[6] that was issued on Saturday, 15 December 2018. It is contained in decision 8/CMA.1[7] that was read from the podium as opposed to the other parts of the Katowice Rulebook that were posted on the UNFCCC website.[8]

Paragraphs 3 and 4 of decision 8/CMA.1 state that the CMA:

Requests the Subsidiary Body for Scientific and Technological Advice to continue consideration of the mandates referred to in paragraph 1 above, taking into consideration the draft decision texts referred to in paragraphs 1 and 2 above, with a view to forwarding a draft decision for consideration and adoption by the Conference of the Parties serving as the meeting of the Parties to the Paris Agreement at its second session (November 2019);

Notes that information provided in a structured summary referred to in decision 18/CMA.1, paragraph 77(d) is without prejudice to the outcomes on these matters.

This decision for Article 6 is largely procedural, but nevertheless contains two important elements. It refers to two documents, which are "noted." This means that these documents will have standing in the subsequent SBSTA discussions on Article 6 (footnote 8).

The Article 6 part of the Katowice text refers to paragraph 77(d) of the Paris Agreement rulebook, which is part of the Transparency Framework contained in decision 18/CMA.1:[9]

Each Party that participates in cooperative approaches that involve the use of internationally transferred mitigation outcomes towards an NDC under Article 4, or authorizes the use of mitigation outcomes for international mitigation purposes other than achievement of its NDC shall also provide the following information in the structured summary consistently with relevant decisions adopted by the CMA on Article 6:

  (i)   The annual level of anthropogenic emissions by sources and removals by sinks covered by the NDC on an annual basis reported biennially;

  (ii) An emissions balance reflecting the level of anthropogenic emissions by sources and removals by sinks covered by its NDC adjusted on the basis of corresponding adjustments undertaken by effecting an addition for internationally transferred mitigation outcomes first-transferred/transferred and a subtraction for internationally

---

6   UNFCCC. Katowice climate package. Available at: https://unfccc.int/process-and-meetings/the-paris-agreement/
    paris-agreement-work-programme/katowice-climate-package.

7   UNFCCC. Decision 8/CMA.1. Matters relating to Article 6 of the Paris Agreement and paragraphs 36–40 of decision
    1/CP.21. FCCC/PA/CMA/2018/3/Add.1. Available at: https://unfccc.int/sites/default/files/resource/cma2018_3_
    add1_advance.pdf#page=22.

8   European Roundtable on Climate Change and Sustainable Transition (ERCST). 2019. Rulebook for Article 6 in the
    Paris Agreement: Takeaway from the COP24 outcome. https://ercst.org/wp-content/uploads/2019/02/Rulebook-
    for-Article-6-in-the-Paris-Agreement-Takeaway-from-the-COP-24-outcome.pdf.

9   UNFCCC. FCCC/PA/CMA/2018/3/Add.2. Available at: https://unfccc.int/sites/default/files/resource/cma2018_3_
    add2_new_advance.pdf.

transferred mitigation outcomes used/acquired, consistent with decisions adopted by the CMA on Article 6;

(iii) Any other information consistent with decisions adopted by the CMA on reporting under Article 6;

Information on how each cooperative approach promotes sustainable development; and ensures environmental integrity and transparency, including in governance; and applies robust accounting to ensure inter alia the avoidance of double counting, consistent with decisions adopted by the CMA on Article 6.

This reference is also explicit on the fact that paragraph 77(d) was agreed in Katowice "without prejudice" to the final outcome of Article 6 negotiations. This could reflect that there is an expectation that the provisions stated in paragraph 77(d) could undergo modifications as part of negotiations for the rulebook for Article 6 (footnote 8).

One interpretation of paragraph 77(d), introduced at the last minute by Parties, is that it outlines an approach needed to avoid double counting under Article 6.2. It could be seen as vague, although it makes the obligatory reference to "consistent with decisions adopted by the CMA on Article 6." However, it can be argued that paragraph 77(d) is about reporting and should not be interpreted as providing any direction for accounting under single-year or multiyear NDC scenarios.

In the absence of further decisions relating to the Article 6 rulebook, could these provisions become the framework for reporting that will ensure that the provisions of Article 6.2 are met in principle? In other words, if there is nothing else, can paragraph 77(d) provide sufficient guidance? If there is no agreement on the Article 6 rulebook, can Parties start developing their own approaches, based on paragraph 77(d)? This may lead to a lack of urgency for some to negotiate on Article 6 at future COPs.

Whether 77(d) would be sufficient or not is debatable. One view, which is not shared by all, is that there may be a high-level guidance framework for avoiding double counting under 77(d). However, in this view, there would still be need for significant documentation and more specific guidance for this to be considered good enough for avoidance of double counting. Further, the use of cooperative approaches that involves international transfer of ITMOs must follow a sound accounting methodology that 77(d) may not cover. This would have to be developed under Article 6.2.

The Katowice outcome for Article 6 raises a second question: whether Parties were inches away from a solution, or if what was needed was to seriously examine if the Parties plowed the right field, need to retrace steps, and go back to fundamentals.

Some may be of the view that Parties were inches away and a final push would have resulted in an agreement being reached. However, with all eyes on one country and the issues it raised in Katowice, it could be tempting to conclude that the matter could be addressed or forced through, and a positive outcome ensue.

That may or may not be true, as there were other Parties who quietly shared that position, even if tactically. Perhaps even more important, there were other issues that were of great concern that stayed out of the limelight of the last-minute discussions, which does not mean they were not present.

Other issues were identified as being significant and they could be summarized as (footnote 8):

(i)   metrics, which is the issue of what ITMOs can be denominated in;
(ii)   corresponding adjustments that will be done either at each transfer or at time of use;
(iii) double counting and the relationship between Articles 6.2 and 6.4;
(iv)   accounting for single-year and multiyear NDCs;
(v)   scope of NDCs or the "inside/outside" debate; and
(vi)   the legacy of Kyoto Protocol mechanisms.

There are two issues that need to be addressed to have a successful conclusion to Article 6.2, according to some. However, their legitimacy in Article 6.2 itself, not a solution, is being contested:

(i)   overall mitigation in global emissions and
(ii)   share of proceeds.

The following six questions, put by the two ministers charged with leading Article 6 (Minister James Shaw of New Zealand and Minister Carolina Schmidt of Chile) during the consultation they had with Groups, serve to illustrate the issues facing Parties in the final moments of COP24 (footnote 8):

(i)   Can you agree to the following landing zone for 6.2: that no corresponding adjustment is required outside NDC until 2031, after which, corresponding adjustments are required?
(ii)   Can you agree to the following landing zone for 6.4: that activities can be inside and outside, but that outside will be correspondingly adjusted only after 2031?
(iii) Can you agree that CDM project transition will be time limited to 2023 and require certain conditions so that the project takes into consideration the NDC?
(iv)   Can you agree to silence on the use of pre-2020 units?
(v)   Can you agree that for use other than for NDCs, if it comes from outside the NDC, there will be a requirement to make a corresponding adjustment after 2023?
(vi)   What other steps are needed to tighten this package from the perspective of environmental integrity?

1.4.2 After Katowice: The Road to Santiago/Madrid

Negotiations and discussions continued in formal (SBSTA) and informal settings after Katowice with Parties trying to focus discussions on the issues that some perceived were "not agreed" in Katowice. There was increased involvement of heads of delegation and even ministers in trying to identify a way forward.

The main school of thought was that the Parties had come very close in Katowice and there were a few intractable issues that needed to be addressed and that all that was needed was to focus on those. Some of them are listed in the previous section.

A different school of thought was that there were substantial differences between Parties on issues of principle and what was visible was just the tip of the iceberg. In this vision, identifying the "issues of principle" or "political issues" was needed before the technical issues that cascade from them could be addressed.

With a new text coming out of SBSTA 50 in Bonn in June 2019, there was hope that at COP25 in Santiago the issue could be addressed, and Article 6 could be incorporated in a completed Paris Agreement rulebook. The content and negotiating process was more oriented the view that Parties had come close in Katowice, rather than the view that there were substantial differences.

1.4.3 Outcome from the 25th Session of the Conference of the Parties

COP25 did not take place in Santiago but rather in Madrid.[10] The discussion again ended up in stalemate and for the second COP in a row there was no decision on the Article 6 rulebook.

At the end of SBSTA 51 there was no positive outcome, that is no decision, so the process was handed over to the Chilean Presidency which produced three texts trying to find a landing that would accommodate everyone's needs. In the end, according to decision 9/CMA.2,[11] the issue was again sent over to SBSTA for further work at SBSTA 52, which was due to take place in June 2020 in Bonn. Due to the measures adopted by Parties and the UNFCCC in response to the coronavirus disease (COVID-19), SBSTA 52 did not take place and COP26 was also postponed. The negotiating process is now awaiting the re-start of negotiations, which may be expected in June 2021.

In decision 9/CMA.2 (footnote 11), the CMA:

1. Notes the draft decision texts on matters relating to Article 6 of the Paris Agreement prepared by the President of the Conference of the Parties serving as the meeting of the Parties to the Paris Agreement at its second session, while recognizing that these draft texts do not represent a consensus among Parties;
2. Requests the Subsidiary Body for Scientific and Technological Advice to continue consideration of the matters referred to in paragraph 1 above at its fifty-second session (June 2020) on the basis of the draft decision texts referred to in paragraph 1 above, with a view to recommending draft decisions for consideration and adoption by the Conference of the Parties serving as the meeting of the Parties to the Paris Agreement at its third session (November 2020).

---

[10]   Less than 2 months before COP25 was set to take place in Santiago, Chilean President Sebastian Piñera announced their withdrawal as COP host, due to political unrest in the country. The Spanish government offered to host the talks instead. https://unfccc.int/news/information-update-on-chile-cop25-to-be-held-in-madrid-on-2-13-december-2019

[11]   UNFCCC. FCCC/PA/CMA/2019/6/Add.1. Available at: https://unfccc.int/sites/default/files/resource/cma2019_06a01E.pdf

In the aftermath of this session, the feeling is again, among a significant number of Parties, that progress had been made, and that only a set of issues remain to be solved. Other Parties, which, while less numerous, nevertheless represent influential Parties and groupings, have made it clear that significant issues remain on the table and these issues cascade down to a significantly large number of issues that can be seen as technical. This group, which was ad hoc at COP24 has become more organized and vocal at COP25 and has managed to coordinate between Parties that traditionally may not have been on the same page on many issues.

This clear difference in perception is certainly troubling, as to solve a problem, it needs to be identified and recognized before it can be solved. The fact that Parties were not able to agree on one text to forward to the next SBSTA session may indicate a lack of convergence. It also shows a desire to go back to an earlier stage where more options were on the table and when the text was seen as owned by Parties, as opposed to the later versions where some options were deleted by the Presidency in its efforts to seek consensus.

While certainly the debate on Article 6 has matured at COP25, the "visible" issues remain the same: corresponding adjustment at Article 6.4 issuance, CDM transition, certified emissions reduction (CER) transition, share of proceeds, and overall mitigation of global emissions. Others, such as registries and baselines and additionality, to enumerate two for illustration purposes, are also in play and may require clarifications and better language but are seen as more technical and also more doable through grinding text negotiations.

1.5 Key Concepts of the Article 6 Rulebook

1.5.1 Nationally Determined Contribution Definition in Relation to Article 6

While this should not be the case, there is a strong debate under Article 6 of what an NDC is. Given that ITMOs and Article 6.4 units (A6.4U) are used toward an NDC, and some of the debate is whether they can be issued from inside and outside the NDC, the definition of an NDC has a significant importance.

One interpretation endorsed by an overwhelming majority is the "classic" one, what has been pledged by a Party, e.g., −40% in emissions by 2030. It would be defined by sectors of the economy covered in the NDC.

An alternative definition, which has emerged in negotiations, is that the NDC is the sum of actions that need to be undertaken to meet the pledge in the NDC, i.e., the NDC is not the pledge, but the actions to accomplish that pledge.

1.5.2 Governance

Articles 6.2 and 6.4 can be described and differentiated in many ways, but the fundamental difference that is built into the Paris Agreement by those who drafted it is that of governance.

Both articles provide the functions that ensure that mitigation outcomes transferred internationally can be used toward the NDC of the Party other than the Party where the mitigation outcome takes place. However, the governance is very different.

Under Article 6.2, the governance is largely, some would say overwhelmingly, non-multilateral or bilateral/plurilateral. That is, the cooperating Parties (note that cooperation is undefined so the simple transfer can be construed as representing cooperation) will define what constitutes an acceptable mitigation outcome to both Parties, how it is measured, certified, what kind of assurance may be acceptable to both Parties that a corresponding adjustment will be undertaken, etc.

The only involvement of the multilateral process is defining what is an acceptable way to show "robust" accounting, including the avoidance of double counting, as well as the rules of transparency in reporting information on the mitigation outcome as well as the transfer and associated actions.

Therefore, the governance under Article 6.2 is mainly decentralized with some elements of centralization to ensure accurate accounting.

On the other hand, Article 6.4 is clearly under a centralized approach as the creation of mitigation outcomes is done under the eye of a supervisory body that is appointed by the CMA. That makes it similar to the CDM where all decisions were taken by the CDM Executive Board and associated bodies (Methodology Panel, Accreditation Panel). As under the CDM (and as under Article 6.3 for Article 6.2 transfers), the discussions in negotiations foresee that the issuing Party under Article 6.4 needs to approve and provide a certification that it approves the transfer.

However, the draft negotiating text under Article 6.4 provides that certain functions could be effectively devolved to the Parties (e.g., baseline definition), under guidelines for the supervisory body. There is, therefore, a stronger element of decentralization under Article 6.4 than under the CDM.

The contrast in governance, as intended by the Parties in Paris, is evident. In devising the rulebook, Parties should resist the temptation to force these two approaches to converge. They will likely converge in time in many practical aspects (e.g., methodologies), but that will be a gradual and natural development.

1.5.3 Environmental Integrity

Environmental integrity is recognized as an important item under Article 6 (it is one of the three "must" conditions under Article 6.2). It is also one of those concepts that are difficult to argue against, but which at the same time, are not defined.

There are different interpretations of environmental integrity, but one that could be considered as having broad acceptance is that a transfer does not lead to an increase in global emissions.

The starting point for ensuring environmental integrity is the stringency of the NDC of the issuing country based on whether targeted GHG emissions are equal or lower than what would be expected under business-as-usual (BAU) conditions.

When the issuing country's NDC is more stringent than BAU, for mitigation activities that are inside the scope of the NDC, the environmental integrity of mitigation outcomes is ensured as long as the issuing country applies the corresponding adjustment to its NDC.

When the mitigation activity is outside the scope of the NDC, the issuing country as well as the acquiring party may want to ensure the unit quality of its mitigation outcome, and the issuing country can also apply corresponding adjustments if it wishes to demonstrate increased ambition of climate action. The term "unit quality" is used widely and interpreted differently in carbon markets. Unit quality refers to the level of confidence that an internationally transferred emissions unit is associated with at least 1 ton of carbon dioxide equivalent (CO_2e) emission reductions, or that it is the result of an actual effort to reduce a ton through the activity.

In cases where the NDC is less stringent than BAU, mitigation outcomes generated from the mitigation activity will require corresponding adjustments to the issuing country's NDC as well as additional measures to ensure environmental integrity through the assessment of the unit quality of the mitigation outcome.

There are different ways to ensure unit quality and different levels of crediting baselines that countries can adopt. The approach for unit quality can use discounting of the mitigation outcome to address uncertainties related to the baseline setting. Among the possibilities, the discounting can be based on the relative mitigation value, which can be determined by the relative ambition of NDC targets in the two trading countries.

Another consideration in determining environmental integrity is increasing the climate ambition of the country and scope of the NDC targets over time to address concerns with weakening climate ambition caused by the transfer of mitigation outcomes.

1.5.4 Sustainable Development

Sustainable development is another one of the "must" conditions under Article 6.2, and like environmental integrity, undefined. However, it is more contentious and elicits stronger reactions.

Traditionally, sustainable development has been seen as a typical one-size-does-not-fit-all and the national prerogative of Parties. Most Parties have strongly resisted any calls to define it at the international level. In the CDM, a simple certification from the designated national authority that the project activity met the sustainable development priorities of the country was sufficient. However, a voluntary tool to define sustainable development was provided in the CDM, over the objections of some Parties.

The situation is not dissimilar in Article 6 where there is much talk about sustainable development, but no significant draft text to support its operationalization.

Sustainable development can be seen as featuring prominently in the preamble of Article 6 as a unitary objective together with raising climate action ambition and ensuring environmental integrity.

Article 6.2 states that Parties can decide to enter into voluntary cooperation and transfer mitigation outcomes and in doing so "Parties shall […] promote sustainable development and ensure environmental integrity."

In Articles 6.4 to 6.7, "a mechanism to contribute to mitigation and support sustainable development" is established.

Articles 6.8 and 6.9 define a "framework for nonmarket approaches to sustainable development."

Yet, without clear guidance and rules on how to promote sustainable development, there is a risk of repeating the CDM's failure to deliver tangible sustainable development contributions.[12] Building on the experience and lessons learned from the sustainable development assessment in the Kyoto Protocol and voluntary market mechanisms, the challenge is how the sustainable development provisions in Article 6 can be operationalized to incentivize a "race to the top" as opposed to a "race to the bottom."[13]

Since the urgency of a transition toward sustainable development and net zero global GHG emissions was underlined in the Special Report on Global Warming of 1.5°C,[14] the concept of transformation has gained momentum particularly in the climate finance community.

Advancing the concept of transformative Article 6 activity design promotes the achievement of Sustainable Development Goals (SDGs), allows for NDC ambition-raising and complements the additionality criteria to safeguard the environmental integrity of cooperative approaches.

Transformational change and its integration into climate change mitigation activities in effect mainstream sustainable development, the former so-called co-benefits of mitigation activities, into outcomes at scale contributing to net zero emissions by 2050.

Simply explained, the concept of transformative impact is the result of climate and sustainable development outcomes at scale, sustained over time. A more elaborate definition and methodology to assess transformational impact of policies and actions is described in the Initiative for Climate Action Transparency Transformational Change Methodology. The assessment of transformative impacts of Article 6 programs to enhance the ambition of NDC implementation enables that synergies for climate and sustainable development are promoted and negative trade-offs are mitigated or avoided. Advancing

---

[12]   K. H. Olsen. 2007. The Clean Development Mechanism's Contribution to Sustainable Development: A Review of the Literature. *Climatic Change*. 84 (1). pp. 59–73; CDM Policy Dialogue. 2012. Climate Change, Carbon Markets and the CDM: A Call to Action Report of the High-Level Panel on the CDM Policy Dialogue. Bangkok: UNFCCC.

[13]   C. Sutter and J. C. Parreño. 2007. Does the Current Clean Development Mechanism (CDM) Deliver its Sustainable Development Aim? An Analysis of Officially Registered CDM Projects. *Climatic Change*. 84 (1). pp. 75–90.

[14]   IPCC. 2018. Global Warming of 1.5°C: An IPCC Special Report on the Impacts of Global Warming of 1.5°C above Pre-Industrial Levels and Related Global Greenhouse Gas Emission Pathways, in the Context of Strengthening the Global Response to the Threat of Climate Change. In Press.

the concept of transformative Article 6 activity design promotes the achievement of the SDGs, allows for NDC ambition-raising and complements the additionality criteria to safeguard the environmental integrity of cooperative approaches.

1.5.5 Double Counting

Double counting is one of the concepts that have, surprisingly, proven to be a significant stumbling block in the effort to finalize the Article 6 rulebook. In principle, double counting, or the avoidance of double counting, takes us to the simple approach of double-entry accounting, which has been around for a long while.

One can argue that the discussion can be focused in a number of ways. One approach is to focus on the type of "double" we are referring to: issuance and usage. Another approach is to focus on what is counted, and while a third approach is what it is counted toward.

It is quite clear that double counting both at issuance and at usage is something that needs to be avoided at all costs if the credibility of actions under Article 6 is to endure.

The second issue is what is being counted. In this case, the issue has been cast as whether ITMOs from inside the NDC only, or both inside and outside the NDC should be accounted for. Accounting for these ITMOs (and avoiding double counting) has led to disputes on the definition of an NDC, which was discussed in Chapter 1.5.1.

In Article 6.2, there is clear reference to that fact that a transfer needs to be accounted for and a corresponding adjustment to be undertaken when a transfer takes place (including in para. 36 of 1/CP.21). This has been interpreted by many as implying that a transfer in Article 6.2 can only take place from under the NDC, and that there needs to be a corresponding adjustment.

However, under pressure from Parties, there is reference under Article 6.2 for accounting and avoidance of double counting of ITMOs transferred from outside the NDC, with the condition that double counting is avoided by also undertaking a corresponding adjustment. The logic is not easy to find as to how one can adjust an NDC if the ITMO comes from outside the NDC.

Under Article 6.4, nowhere in the text is there a reference to the avoidance of double counting through corresponding adjustments. This is interpreted by a group of Parties as implying that there need not be a corresponding adjustment at first issuance under Article 6.4, but only from second transfer on. They accept that there cannot be double counting and point to the language in Article 6.5 which has different provisions, but no reference to corresponding adjustments.

This leads directly to the third issue that we need to focus on, and which has deeply divided Parties: avoiding of double counting toward what—emissions (or inventory balance) or NDC.

It must be emphasized that para. 36 of 1/CP.21 (footnote 5) makes the following reference to emissions:

[...] ensure that double counting is avoided on the basis of a corresponding adjustment by Parties for both anthropogenic emissions by sources and removals by sinks covered by their nationally determined contributions under the Agreement.

Articles 6.2–6.3 and Articles 6.4–6.5 refer to how ITMOs are to be used "towards meeting [NDCs]"

Article 6.2:

"[...] that involve the use of internationally transferred mitigation outcomes towards nationally determined contributions [...]

Article 6.3:

"The use of internationally transferred mitigation outcomes to achieve nationally determined contributions under this Agreement shall be voluntary and authorized by participating Parties."

Article 6.4:

"[...] resulting in emission reductions that can also be used by another Party to fulfil its nationally determined contribution."

Article 6.5:

"Emission reductions [...] shall not be used to demonstrate achievement of the host Party's nationally determined contribution if used by another Party to demonstrate achievement of its nationally determined contribution."

It is not illogical to conclude that the avoidance of double counting needs to be toward NDC accounting. NDCs and NDC accounting represent the fundamentals of the Party commitment and is critical for the global stocktake. This does not in any way detract from the need to accurately report inventories or inventory balance, adjusted for transfers of ITMOs, to have an accurate global picture.

This dispute has led to the debate regarding para. 77(d) under the transparency framework (footnote 9), which refers to emissions but makes no reference whatsoever to accounting (and therefore the avoidance of double counting) toward NDCs.

Paragraph 77 (d) states that Each Party that participates in cooperative approaches that involve the use of internationally transferred mitigation outcomes towards an NDC [...] shall also provide the following information in the structured summary consistently with relevant decisions adopted by the CMA on Article 6.

It is interesting, and important, to note that reference is made to NDC and use "towards an NDC", but the rest of this paragraph talks about emissions balance. They could be the same, but they are not necessarily the same. The text seems to ignore this fact.

1.5.6 Relationship between Articles 6.2 and 6.4

As mentioned above, Articles 6.2 and 6.4 were written with the intention to provide, to some degree, the same function—ensuring that mitigation outcomes can be used toward the NDC by a Party other than the one where the mitigation outcome took place—that is, mitigation outcomes can be transferred internationally.

In addition, Article 6.4 is intended to provide a protocol, under the Supervisory Board appointed by the CMA that will certify these mitigation outcomes.

In terms of the relationship between Articles 6.2 and 6.4, the aspect most stakeholders will dispute is the relationship between an Article 6.4 emissions reductions units (A6.4U) and the ITMOs, which is what exists in the Article 6.2 universe. It is clear that A6.4U transfers need to be accounted for as well in terms of meeting NDCs. It does not strike as logical that there should be two totally separate accounting systems: one for Article 6.2 and another for Article 6.4.

 The question then: Is an A6.4U an ITMO? When does an A6.4U become an ITMO: at issuance or at some point in its existence; between when it is issued into the A6.4 Registry (A6.4R) or when it is used toward an NDC?

There are some Parties, the majority of whom will make the case that an A6.4U becomes an ITMO as soon as it reaches the receiving Party after the first transfer—an A6.4U is issued into the central registry, after which it is transferred to the purchasing Party, which becomes the first receiving Party.

It is unclear to the author why the initial issuance needs to take place in a central A6.4R, instead of the issuing Party registry. There was a rationale for that under the CDM when the host Party did not have a registry, but that is no longer the case—version 3 of the Madrid draft text states that under Article 6 of the Paris Agreement all Parties should have a Registry.

The reasoning could be that a receiving Party needs to make a corresponding adjustment and therefore, the issuing Party should also make one to maintain the double entry accounting concept intact. If an A6.4U is issued in a central A6R then one may make the argument that it is not deemed to be an international transfer.

The other school of thought is that an A6.4U becomes an ITMO after the second transfer, i.e., when the first receiving Party makes a transfer to a second receiving Party. The rationale in this case is that there is no reference to a corresponding adjustment under Article 6.4, and that the avoidance of double counting is elaborated in Article 6.5 without any reference to corresponding adjustments.

The argument also goes that any A6.4U would not be issued from the NDC, but would be done from an overachievement of the NDC—what is achieved in addition to meeting the NDC. Therefore, there is no need to do a corresponding adjustment to the NDC. This of course assumes that the definition of the NDC is not the "classic" one, i.e., what has been pledged by a Party, e.g., −40% in emissions by 2030, as outlined in Chapter 1.5.1.

2. Articles 6.2 and 6.3: Negotiation Issues

This section identifies ongoing negotiation issues concerning Article 6.2 and 6.3 and gives a detailed account of the main points of contention.

The identified issues are related to definitions, participation, corresponding adjustments, sectors and gases, other international mitigation purposes, limits to the transfer and use of ITMOs, reporting and review under Article 6, infrastructure for recording and tracking of ITMOs, and ambition in mitigation and adaptation actions, all as contained in the version 3 of the draft text proposed by the President.

2.1 Definitions

The Definitions section of the rulebook is always a contentious one from a process and substance point of view. While some Parties find it necessary, others argue that it introduces a significant amount of contentious issues that would be better left defined in the operational parts of the rulebook. Those who want to have this section argue that providing definitions through operational elements will lead to a lack of clarity and consistency, and leave many issues open to interpretation, which they wish to avoid.

The definition of ITMOs contained in the proposal by the President includes a number of items starting with "Real, verified, and additional."[15]

In this case, reference is made to real and verified, which no one can take issue with. The only issue that has raised some questions is the reference to additionality, normally associated with mitigation outcomes emerging from baseline-and-credit mechanisms (such as the CDM). This may eliminate any mitigation outcome from cap-and-trade systems, which many Parties may object to. However, some have argued that scarcity in a cap-and-trade system is equivalent to additionality in the case of baseline-and-credit systems. If a cap is determined to be too high, there could be more allowances than are required in a BAU scenario so that the end result is the same. BAU and post-cap scenarios would be the same.

---

[15]   Refer to paragraph 1(a) of the Draft decision texts on guidance on cooperative approaches referred to in Article 6, paragraph 2, of the Paris Agreement (third iteration, 15 December), available at https://unfccc.int/documents/204687.

Should this provision remain in the final text, it could be interpreted as not allowing linking of emission trading systems (ETSs) under Article 6.2, which would potentially be a serious drawback.

This debate is especially important when addressing the issue of additionality and its relevance for baseline-and-credit and cap-and-trade systems. Another important aspect could be related to the timing and form of corresponding adjustment, i.e., emissions trading could be adjusted through a netting provision, ex-post, whereas baseline-and-credit transfers should be adjusted at each transfer.

Another element included in the definition of ITMOs is:

[...] including mitigation co-benefits resulting from adaptation actions and/or economic diversification plans, or the means to achieve them.[16]

This provision is an important point for some Parties that focus on adaptation and economic diversification in their NDC and feel that any mitigation outcome from these actions should qualify as an ITMO.

Removals is clearly mentioned in this version of the text but no explicit reference to what kind of removals it is referred to. There is general acceptance of sinks, except a natural sinks provision (nature-based solutions as opposed to technology driven approaches such as carbon capture and storage), which is seen by some Parties as exclusively belonging to Article 5 (forests, REDD+[17]) of the Paris Agreement. Some Parties have indicated that including REDD+ as an eligible activity under Article 6 could be a red line for them. This may change as domestic dynamics in some Parties change.

Excluding avoidance from the definition has raised concerns from some Parties. The reason for exclusion is the lack of definition (some still fear the re-emergence of the definition related to the avoidance of producing hydrocarbons), as well as the definition of the baseline, which is seen as potentially complex to manage.

Measured in metric tonnes of carbon dioxide equivalent [tCO_2e] in accordance with the methodologies and common metrics assessed by the [Intergovernmental Panel on Climate Change] and adopted by the CMA [and/or in other non-greenhouse gas metrics determined by participating Parties [that are consistent with the nationally determined contributions (NDCs) of the participating Parties]].[18]

The metrics issue has been a challenging one and can be seen as one of the root issues, which cascades into other issues; some of principle, others of operational nature. This was one of the issues that deadlocked COP24: how to deal with the demand by some Parties for ITMOs in metrics other than CO_2e—is there a need for them, can they be operationalized, and does that lead to other outcomes, such as the need for a buffer or netting account(s)?

---

[16]   Footnote 15, Paragraph 1 (b).
[17]   Reducing emissions from deforestation and forest degradation. https://unfccc.int/topics/land-use/workstreams/redd/what-is-redd.
[18]   Footnote 15, Paragraph 1 (c).

Metrics have emerged from two core issues:

(i)   Some Parties would like to be able to sell ITMOs in metrics that are relevant to the transaction they are undertaking and relevant to the baseline in their country.

(ii)   Other Parties feel that if ITMOs are all in CO_2e, then any Party participating in Article 6 would be obliged to express their NDC in CO_2e, and also show progress toward their NDC in that metric. This clashes with their view that the Paris Agreement is a bottom-up agreement, and Parties are allowed to choose their own NDC metrics.

Some Parties object to a CO_2e metric as a matter of principle: Do the NDCs have to adapt to the accounting rules of Article 6 or do the Article 6 rules need to be inclusive?

The bottom-up conception of NDCs seems to be a fundamental logic and it might be difficult to find a way of marketing ITMOs in the "local currency" without forcing Parties to express NDCs in CO_2e. The biggest challenge would be to find the conversion factor that would allow Parties to exchange between the non-CO_2e-denominated ITMOs and those expressed directly in CO_2e.

The other side of the argument is that using one metric (CO_2e) would allow for transparency and comparability and help with market liquidity. Ease of working for the 2023 Global Stocktake mandated by the Paris Agreement is also a consideration in pushing for a CO_2e-only agreement.

The argument is interrelated with the other issues under discussion. Metrics will impact what is being adjusted, whether it is the netting account and then the NDC, or the emissions related number. Also, more than one metric would, in principle, require an equivalent number of netting or buffer accounts to accommodate that.

It is important to understand the buffer or netting account as an account set initially at 0 and which is adjusted every time there is a transfer, in the respective metric of that netting account. One could argue that it is simply providing the ability to keep account of ITMOs in different metrics, and it is adjusted. There is a difference between adjustment (to a netting account) and corresponding adjustment (to an NDC), an issue that has been highlighted many times in negotiations by the proposing Parties, but which some Parties refuse to contemplate.

The idea is that at the end of each year, the number in the netting account is used to do the Paris Agreement-mandated corresponding adjustment, in the metric of the Party's NDC. If ITMOs are used from an account in a different metric, then a conversion factor will have to be used. Clearly, the issue of how to determine the conversion factors, as well as their governance, will need to be solved at future sessions of SBSTA.

Highlighting the difference between adjustment and corresponding adjustment is important.

As well as accounting, the metrics that ITMOs are measured in are also a matter of principle, but would seem to rather flow from the accounting issue, discussed in

Chapter 1.5.5. What are we accounting for: NDC accounting or emissions accounting? This is the defining issue for the Article 6.2 rulebook.

The provision included in the last sentence *("consistent with the nationally determined contributions (NDCs) of the participating Parties")* would allow for ITMOs transferred in other metrics, provided that the two NDC metrics of the participating Parties in cooperative agreements match. This is seen by some Parties as limiting the scope of such exchanges (which is for them desirable), while at the same time eliminating the need for any conversion factors. Others, who do not approve of this provision, see it as an unjustified limitation put on the transfer of non-CO_2e-denominated ITMOs.

From a cooperative approach referred to in Article 6, paragraph 2 of the Paris Agreement, (hereinafter referred to as a cooperative approach) that involves the international transfer of mitigation outcomes authorized for use towards an NDC pursuant to Article 6, paragraph 3 of the Paris Agreement;[19]

The issue of a cooperative approach having to include two or more Parties from inception emerged leading in to SBSTA 50, as it was not on the radar at COP24. It is essentially a repetition of the discussion on unilateral CDM, whereby units could be created by a project owned by a developing country project developer, to be sold at a time of its choice to a customer of his or her choosing.

Some see the requirement of including a second Party from the outset not only as a constraint, but also as a possible interference in the NDC. If ITMOs need to be created bilaterally, then the mitigation outcomes behind the ITMOs cease to be a national prerogative and are subject to international, or, if not, at least bilateral governance. This could imply that the cap on ETSs around the world become subject to some level of international interference under the prospect of linking. While some Parties will see this as necessary, others will see this going completely against the bottom–up ethos of the Paris Agreement.

It is unclear if both receiving and issuing Parties need to actually provide an Article 6.3 certification. This would make the difference between unilateral ITMOs and bilateral ITMOs, an issue that has been present in the CDM debate for a long time.

Mitigation outcomes authorized by a participating Party for use for international mitigation purposes other than achievement of its NDC or for other purposes, including as determined by the first transferring participating Party (hereinafter referred to as other international mitigation purposes).[20]

This provision relates to the desire of many Parties to be able to count ITMOs toward International Civil Aviation Organization (ICAO) and possible International Maritime Organization compliance use.

---

[19]   Footnote 15, Paragraph 1 (d).
[20]   Footnote 15, Paragraph 1 (f).

This provision is being contested by a number of Parties that argue that ITMOs, according to Article 6, are to be used toward NDC and not for other purposes:

6.4ERs under the mechanism established by Article 6, paragraph 4 when they are internationally transferred.[21]

This provision is intended to address the relationship between A6.4U and ITMOs, discussed above in Chapter 3.4.5. However, as currently formulated, it would not provide much help as internationally transferred can easily be interpreted as referring to the transfer between the first receiving Party and the next transfer-in Party, i.e., at second transfer.

2.2 Participation

Participation responsibilities are not something new as eligibility requirements existed for participation in Articles 6, 12, and 17 (markets articles) under the Kyoto Protocol (see Box 2).

The provision about ratification of the Paris Agreement cannot be controversial as Parties are expected to be Parties to the Paris Agreement to use Article 6.

Box 2: Eligibility Criteria for the Kyoto Protocol

To participate in the Kyoto Protocol mechanisms, Annex I Parties must meet the following eligibility requirements:

- They must have ratified the Kyoto Protocol.
- They must have calculated their assigned amount in terms of tons of carbon dioxide (CO_2)-equivalent emissions.
- They must have in place a national system for estimating emissions and removals of greenhouse gases within their territory.
- They must have in place a national registry to record and track the creation and movement of emission reduction units, certified emissions reductions, assigned amount units, and removal units must annually report such information to the secretariat.
- They must annually report information on emissions and removals to the secretariat.

Source: UNFCCC. FCCC/KP/CMP/2005/8/Add.2 Decision 11/CMP.1. Modalities, rules and guidelines for emissions trading under Article 17 of the Kyoto Protocol. Available at: https://unfccc.int/resource/docs/2005/cmp1/eng/08a02.pdf.

The other provisions in this section also refer to actions that Parties would be expected to do if they are to use Article 6. The one issue that may be questionable from a pure Article 6

---

[21]   Footnote 15, Paragraph 1 (b).

perspective is the provision on inventories as Parties are expected to use ITMOs toward the NDC.[22]

However, if consideration is given to how there will be need for accounting on compliance toward the NDC (with regard to Party commitments) and toward the global goal which will be expressed through inventory balance, then it can be argued that this requirement is justifiable.

One provision that would seem to be missing is that Parties need to have a registry. Such a provision may be opposed by those that see the need for a registry only if a Party is a buyer, not in the case where it is a seller. If it is a seller only and uses only Article 6.4, the argument would be that the issuance takes place in the A6.4R and there is no need for a national registry.

2.3 Corresponding Adjustments

This issue is divided into a number of sections, starting with metrics. The issue of metrics is referring to guidance and relevant decisions of the CMA, so a decision was to be avoided in Madrid.

2.3.1 Application of Corresponding Adjustment

The header of paragraph 8 (footnote 15) outlines a number of things that need to be avoided when applying corresponding adjustments:

Each participating Party shall apply corresponding adjustments in a manner that ensures: transparency, accuracy, completeness, comparability and consistency; that participation in cooperative approaches does not lead to a net increase in emissions within and between NDC implementation periods; that corresponding adjustments shall be representative and consistent with the participating Party's NDC implementation and achievement.

While all these issues are clearly desirable, it is also less evident how some of them can be operationalized. There are certain things that Parties will do individually and where one Party cannot possible be held accountable for the actions of the other. The language should be clarified and specific without the joint responsibility that seems to exist from the current draft.

Single-year NDCs are clearly challenging in terms of applying corresponding adjustments, with the solutions being proposed outlined as follows (footnote 15) and some examples in Box 3:

(i)   *Providing a multi-year emissions trajectory, trajectories or budget for the NDC implementation period that is consistent with implementation and achievement of the*

---

[22]   Annual inventories are required from 2020 from all countries. Para 57. Decision 18/CMP.1

NDC, and annually applying corresponding adjustments for the total amount of ITMOs first transferred and used for each year in the NDC implementation period;

(ii)   Calculating the average annual amount of ITMOs first transferred and used over the NDC implementation period, by taking the cumulative amount of ITMOs and dividing by the number of elapsed years in the NDC implementation period and annually applying indicative corresponding adjustments equal to this average amount for each year in the NDC implementation period and applying corresponding adjustments equal to this average amount in the NDC year....[23]

Box 3: Corresponding Adjustments: Methods for Single-Year Targets

The details for the methods that can be used for addressing single-year targets are not yet elaborated. The following visualizations can be used as examples of approaches.

Visualization of corresponding adjustments against a multiyear trajectory

Figure B3.1: Corresponding Adjustment against a Multiyear Trajectory of Seller Country

ITMOs bought/CA made annually

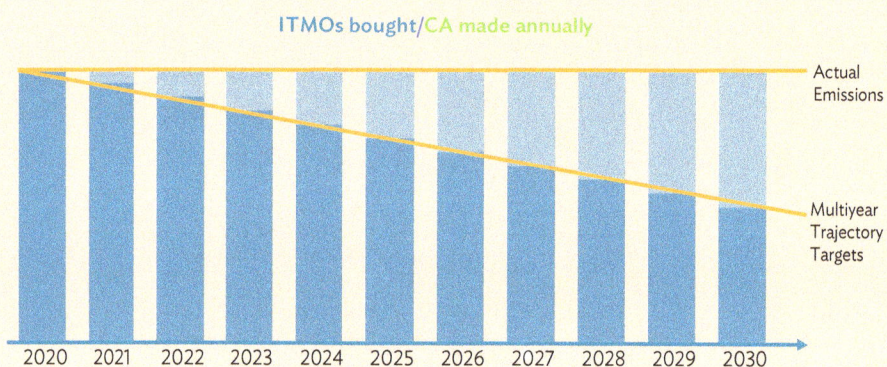

Figure B3.2: Corresponding Adjustment against a Multiyear Trajectory of Buyer Country

ITMOs transferred/CA made annually

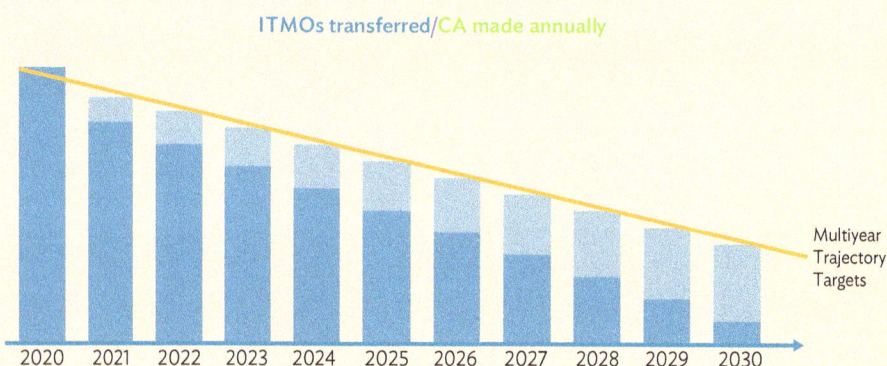

continued on next page

---

[23]   Footnote 15, Paragraph 8 (a) (i) and (ii).

Box 3 continued

Visualization of averaging corresponding adjustments

Figure B3.3: Averaging Corresponding Adjustments by Seller Country

ITMOs bought

CA Made

Target

2020 2021 2022 2023 2024 2025 2026 2027 2028 2029 2030

Figure B3.4: Averaging Corresponding Adjustments by Buyer Country

ITMOs transferred

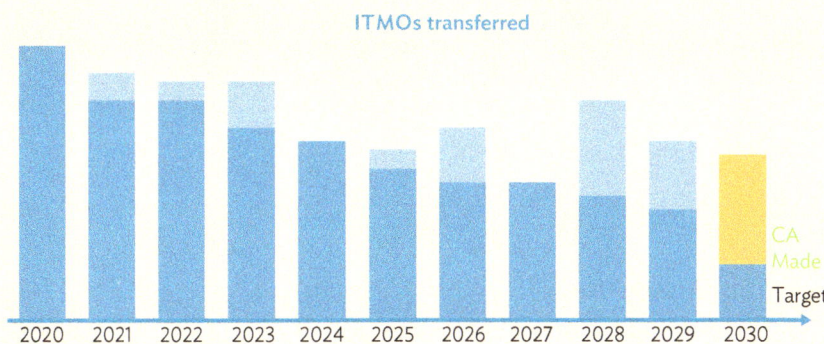

CA Made

Target

2020 2021 2022 2023 2024 2025 2026 2027 2028 2029 2030

CA = corresponding adjustment, ITMO = internationally transferred mitigation outcome.

Source: Adapted from Greiner S., Krämer N., Michaelowa A. and A. Espelage 2019. Article 6 Corresponding Adjustments. Key accounting challenges for Article 6 transfers of mitigation outcomes. Climate Focus, B.V., Perspectives Climate Group GmbH.

Parties not only have NDCs in different metrics, but also in different formats—some express them as multiyear, while others express them in single-year commitments. This is a departure from the Kyoto Protocol where all commitments were expressed in budgets that made accounting easy to understand. The way accounting is done will weigh heavily on investment decisions, as this will impact the economics of projects.

The two options that have provisions in the text are NDCs with single-year and multiyear commitments. The options provided require that whatever method is applied by a Party is used consistently during the NDC period. However, the term "consistency" is still under negotiation, and whether it refers to each cooperating Party, or to all cooperating Parties during the NDC period is yet to be clarified.

Essentially the issue that is being debated is how the transfers during the NDC period would be counted.

(i)   **Calculating a multiyear trajectory**. This is not difficult to implement but some Parties would argue that they would need to express their NDC in a manner other than what they had nationally determined. Under this approach, Parties would also need to apply a corresponding adjustment for each year if there is a multiyear commitment. It is assumed that ITMOs transferred in the respective year would be the only ones used.

(ii)   **Calculating the average.** This involves getting the average of the ITMO transfers over the NDC period (e.g., 100 over 10 years) and applying to each year of the NDC the average amount.

It needs to be recognized that these approaches are somewhat subjective, and they all present advantages and disadvantages. The choice, whether it is admitted or not, will be political.

Objectively, the choice of method should be driven by a number of considerations, some practical, some of principle, including, but not limited to the following:

(i)   How does it relate to the impact on the atmosphere as expressed by the NDC? Parties have chosen to express their contribution to limiting the impact on the atmosphere in different ways and what is accounted should have some relation to the NDC.

(ii)   Which level of liquidity is needed for markets to function well?

(iii) How do you ensure availability of ITMOs in the context of effort of Parties to increase their level of ambition?

One provision that needs to be highlighted is in para. 12 where "a method proposed by a Party that meets the requirements of this chapter III, and this guidance, may be included[...]."[24]

This outcome is the result of some options having been eliminated over the last few sessions and with some Parties wanting to retain the option to have them considered (i.e., cumulative accounting whereby all ITMO imported during the NDC period would be counted toward the NDC).

Another axis for discussion is with respect to corresponding adjustments under different metrics.

Each participating Party with an NDC measured in [tCO₂e] shall apply corresponding adjustments pursuant to paragraph 8 above, resulting in an emissions balance, reported pursuant to paragraph 23 for each year, by applying corresponding adjustments in the following manner to the emissions and removals from the sectors and GHG covered by its NDC:

---

[24]   Footnote 15, Paragraph 12.

(a)   Adding the quantity of ITMOs authorized and first transferred, pursuant to paragraph 8 above;

(b)   Subtracting the quantity of ITMOs used pursuant to paragraph 8 above.

Each participating Party with an NDC measured in non-GHG metrics determined by the participating Parties engaging in a cooperative approach involving ITMOs traded in the same non-GHG metric shall apply corresponding adjustments in a buffer registry by applying an addition to and subtraction from the annual level of the relevant non-GHG indicator used by the Party to track progress towards the implementation and achievement of its NDC in accordance with decision 18/CMA.1 and consistent with this chapter III, this guidance and relevant further guidance of the CMA.[25]

For NDCs measured in CO_2e, the corresponding adjustment is done on the basis of CO_2e, starting from the inventory, without specifying which inventory (in terms of timing) it refers to. One issue that was raised is the difference between the timing of the inventory and that of the corresponding adjustment. Inventories have a time lag and have certain governance, while the timing of corresponding adjustment is different as it is the governance of corresponding adjustment. These differences need to be taken into account.

For NDCs in another metric, the subtraction or addition is made starting from "subtraction from the annual level of the relevant non-GHG indicator used by the Party to track progress [...]."[26] This way of expressing the concept of a netting, interchange, or buffer account (note that the expression "buffer registry" may be somewhat causing confusion) is a new way of presenting this concept. It is certainly a departure from UNFCCC negotiating sessions language and is not representing the concept in an accurate way. It may be causing more confusion as it introduces language that emerged from outside negotiations.

The netting account concept, as conceived, would see the account set at zero (0) at the start of the NDC period for each metric that the Party wishes to transact in, and is updated every time there is a transfer that is authorized as an ITMO.

The netting account at the end of the NDC period represents the net position in that metric for a Party. It is an intermediate number used to adjust the NDC.

The opposition is believed to emerge from some Parties not wanting different metrics (which are associated with netting accounts) as well as an opposition to adjusting NDCs—adjusting emissions is what those Parties wish to have. This is where the discussion of principle that an NDC may be the same as the inventory, but not necessarily, and which was discussed in Section 1.5.1, provides clarity.

When there is a need to show progress toward the NDC, the NDC is adjusted by using the number in that account. There can be multiple netting accounts, one for each currency or metric that the respective Party holds. The account that holds ITMOs in the same metric or currency as the NDC is used directly for the corresponding adjustments. Accounts that hold ITMOs in other metrics will need to be converted into the currency of the NDC, so

---

[25]   Footnote 15, Paragraphs 9 and 10.
[26]   Footnote 15, Paragraph 10.

they can be used in corresponding adjustments. The governance of determining conversion factors has not been discussed and will have to wait for future SBSTA sessions.

This may be easily compared to what everyone is familiar with, a bank account in multiple currencies. One can use money from the bank account in the country directly, while using money from accounts in other currencies will require an exchange rate (in the case of ITMOs, a conversion factor).

2.4 Sectors and Gases

These paragraphs are also referred to as discussing the cases of ITMOs originating from inside or outside the NDC.

A participating Party that first transfers ITMOs from emission reductions and removals from sectors and GHGs covered by its NDC shall apply corresponding adjustments consistent with this guidance.

A participating Party that first transfers ITMOs from emission reductions and removals from sectors and GHGs that are not covered by its NDC shall apply corresponding adjustments consistent with this guidance.[27]

It is clear from Decision 1/CP.21 that any ITMO will have to trigger a corresponding adjustment to avoid double counting. That would seem to imply that ITMOs can be generated only from inside the NDC-covered sectors and gases. If a transfer takes place from sectors and gases not covered by the NDC of the transferring Party, the first reaction would be that it cannot qualify as an ITMO.

Alternatively, if such a transfer were to take place, and was agreed by the CMA that it could be an ITMO, it would then follow that there is no logical reason why it would trigger a corresponding adjustment to the NDC since it is not part of the NDC. The only rationale would be that by transferring a mitigation outcome that had taken place outside the NDC, we are essentially anticipating future adjustments that will inevitably take place if we are to reach zero emissions—the Paris Agreement assumes that Parties will broaden the scope of their NDC over time . It is not an illogical argument, but in the context of accounting under the Paris Agreement, it seems rather tenuous and to complicate things.

---

[27]   Footnote 15, Paragraphs 14 and 15.

2.5 Other International Mitigation Purposes

The draft text is rather cryptic:

Where a participating Party authorizes mitigation outcomes for other international mitigation purposes, it shall apply a corresponding adjustment, consistent with this guidance, for first transfer, whether or not the mitigation outcomes have been internationally transferred.[28]

It is difficult to discern the text as currently drafted. It seems to imply that a corresponding adjustment needs to be undertaken at issuance if there is authorization for use for purposes other than NDC compliance. The text would benefit from being more explicit. If the ITMO is used for non-NDC purposes (e.g., ICAO compliance), there is no formal provision for avoidance of double use between ICAO and Paris Agreement.[29]

Similarly, the text is silent on the use of an ITMO for claims by entities such as corporations who use it for voluntary neutrality, which could lead to the use of ITMOs in multiple claims for their CO_2 characteristics. There is no specific provision on how to avoid double counting for use for other purposes, whatever they may be.

One provision that could be envisaged is some way to ensure coordination for ensuring that the same ITMO cannot be used for multiple purposes. This will likely require measures at the national level as well.

The use of ITMOs for purposes other than NDCs is firmly opposed by a group of Parties whose position is that ITMOs are created for the express purpose of use toward an NDC, as specified in the Paris Agreement. ICAO members, the argument goes, are not Parties to the Paris Agreement and therefore cannot use ITMOs.

2.6 Limits to the Transfer and Use of Internationally Transferred Mitigation Outcomes

This provision is not detailed in the third version of the draft text from Madrid, and it refers to further guidance that is to be developed under the work program that was envisaged following COP25.[30] This issue was not deemed essential to be agreed at COP25 and could wait and not sit in the way of an agreement in Madrid.

[28] Footnote 15, Paragraph 16.
[29] There is a requirement under the Carbon Offsetting and Reduction Scheme for International Aviation (CORSIA) that a host country has to attest to that the emission reduction is not also counted against national targets. If the Article 6 text were explicit about this, then it could be include requiring corresponding adjustments which then would be a method for the attest.
[30] Footnote 15, Paragraph 1 (d).

Limits to transfers are not a novelty in carbon markets, they were present under the Kyoto Protocol in the form of the Commitment Period Reserve (at issuance) and supplementarity (at use). In the case of the Kyoto Protocol, the limitations were, in many respects, easier to define as commitments were expressed as budgets and there were clear ways to express these limitations in relation to budgets.

Given the diverse nature of the NDCs, defining such limitations is not always obvious. At this stage, the current text is very broad and general in nature, simply referring to tracking progress toward NDC, and not leading to an increase in emissions in an NDC period, and between NDC periods.

This issue is generally favored by countries that are looking to ensure that most mitigation actions are undertaken domestically and regard transfers and trading with suspicion. Most developed countries will not be supportive of limits, but would see use in ensuring transparency in reporting progress toward NDCs as a means of avoiding situations of noncompliance later on.

2.7 Reporting and Review under Article 6

In addition to the information related to corresponding adjustments, the enhanced transparency framework (ETF) modalities, procedures and guidelines (MPG) require Parties to report, in its "structure summary" in the biennial transparency report (BTR), "how each cooperative approach promotes sustainable development; and ensures environmental integrity and transparency, including in governance; and applies robust accounting to ensure inter alia the avoidance of double counting."[31] No further guidance is given in the ETF MPG on the level of detail and/or specificity of such information.

Article 6 negotiation draft texts have proposals for additional information requirements to be presented as an initial report; annual information (to be submitted in an agreed electronic format) and regular information to be included in the BTR.

2.7.1 Initial Report Requirements

Time of submission. Information to be presented "no later than the time of providing or receiving authorization or initial first transfer of ITMOs from a cooperative approach and where practical, in conjunction with the next due BTR for the period of NDC implementation."[32]

Content. The following should comprise the initial information:

(i)   demonstration that the participating Party fulfills the participation responsibilities referred to in Chapter II (Participation);

---

[31]   Footnote 9, Paragraph 77 (d) (iv).
[32]   Footnote 15, Paragraph 18..

(ii) description of its NDC (when the participating Party has not yet submitted a BTR);

(iii) ITMO metrics and the method for corresponding adjustments for multiyear or single-year NDCs that will be applied consistently throughout the period of NDC implementation;

(iv) quantification of the Party's mitigation information in its NDC in tCO_2e, including the sectors, sources, GHG, and time periods covered by the NDC; the reference level of emissions and removals for the relevant year or period; and the target level for its NDC, or where this is not possible, the methodology for the quantification of the NDC in tCO_2e; and

(v) quantification of the participating Party's NDC, or that portion of its NDC, in a non-GHG metric determined by each participating Party.

Format and/or place to report the information. The initial information is supposed to be included in the centralized accounting and reporting platform.

2.7.2 Annual Information Requirements

Time of submission. Information is to be presented on an annual basis.[33]

Content. It should contain annual information on ITMO authorization; first transfer; transfer; acquisition; holdings; cancellation; use toward NDCs; authorization of ITMOs for use toward other international mitigation purposes; voluntary cancellation; and specifying the cooperative approach, other international mitigation purposes, first transferring participating Party, using participating Party and vintage, as applicable (footnote 33).

Format and/or place to report the information. The annual information is supposed to be included in the "Article 6 database."

2.7.3 Regular Information Requirements

Time of submission. Information is to be presented on a biannual basis.[34]

Content. The following should comprise the regular information (footnote 34):

(i) demonstration that the participating Party fulfills the participation responsibilities referred to in Chapter II (Participation);

(ii) updates to the information provided in its initial report and any previous BTR;

(iii) information on its authorization(s) of the first transfer and use of ITMOs toward NDCs including any changes to earlier authorizations, pursuant to Article 6, paragraph 3 of the Paris Agreement;

(iv) how corresponding adjustments undertaken in the latest reporting period, pursuant to Chapter III (Corresponding Adjustments) are representative of progress toward implementation and achievement of its NDC; and

---

[33] Footnote 15, Paragraph 20.
[34] Footnote 15, Paragraph 21.

(v)   how it has ensured that ITMOs acquired and used toward achievement of its NDC and those authorized mitigation outcome(s) used for other international mitigation purposes, will not be further transferred, cancelled, or otherwise used.

In addition, regular information includes how each cooperative approach[35]

(i)   contributes to the mitigation of GHG emissions and the implementation of the NDC;

(ii)   ensures environmental integrity, including making sure that there is no net increase in global emissions, through robust, transparent governance and the quality of mitigation outcomes, including through stringent reference levels, baselines set in a conservative way and below BAU emission projections (including by taking into account all existing policies and addressing potential leakage) and minimizing the risk of non-permanence of mitigation and when reversals of emissions removals occur, ensuring that these are addressed in full;

(iii) is measured and transferred in tCO_2e (where the mitigation outcome is in a "GHG metric") in accordance with the methodologies and metrics assessed by the IPCC and adopted by the CMA;

(iv)   is related to other information required by relevant future decisions of the CMA (where a mitigation outcome is measured and transferred in a non-GHG metric);

(v)   provides for, as applicable, the measurement of mitigation co-benefits resulting from adaptation actions and/or economic diversification plans; and

(vi)   applies the limits pursuant to Chapter III.E set out in further guidance from the CMA (Limits to the Transfer and use of ITMOs).

Format and/or place to report the information. The regular information is supposed to be included in the BTR.

According to the proposals, all these additional information are to be reviewed by an "Article 6 technical expert review team" that will produce a "review report," which "may include recommendations to the participating Party on how to improve consistency with this guidance and relevant decisions of the CMA, including on how to address inconsistencies in quantified information."[36] The Article 6 review report is to be forwarded to the technical expert review (TER) under the ETF for consideration. The proposals do not explain what type of "consideration" is to be undertaken during the TER.

Based on similar experience of the technical analysis of REDD+ results submitted as a technical annex of the biennial update reports,[37] it could be expected that the Article 6 technical expert review work in a complementary manner with the ETF TER. Nevertheless, is important to make it clear the different responsibilities of each expert team to avoid overlaps of activities and/or conflicts of assessments. Such clarifications should be included

---

[35]   Footnote 15, Paragraph 22.
[36]   Footnote 15, Paragraph 27.
[37]   More information on the technical analysis of REDD+ results are found in UNFCCC. REDD+ Platform. https://redd.unfccc.int/submissions.html?topic=18; and UNFCCC. Warsaw Framework for REDD-plus. https://unfccc.int/topics/land-use/resources/warsaw-framework-for-redd-plus.

in the Article 6 decisions and in the templates of the TER report[38] and the Article 6 technical expert review report.

2.8 Infrastructure: Recording and Tracking of Internationally Transferred Mitigation Outcomes

The draft text that refers to tracking of ITMOs is somewhat ambiguous and many of the negotiators and stakeholders are to some degree unclear of how to interpret it. Ambiguity is, in many cases, desirable in negotiating texts as it allows Parties to move forward and give their own interpretation to the text.

In this case, it may be driven by how there is no clear and accepted definition of what is an ITMO, leading to difficulties to discuss tracking something that is not yet well-defined.

Article 6 text refers to a number of infrastructure components that are needed to make sure that Article 6 functions properly and positions Parties to meet requirements in Article 6.2, including providing the information needed to ensure integrity, both environmentally and functionally.

Some of the functions that the infrastructure should be able to support include

(i)   ensuring that there is no double counting of ITMOs,
(ii)   making the information available and transparent to Parties and stakeholders to undertake analysis, and
(iii)   tracking ITMOs.

The components mentioned in the latest version of the Article 6.2 text include the following:

(i)   **Registries that each Party has or has access to for the purpose of tracking ITMOs**. It is not directly mentioned in the current text what exactly the registries will track, but one assumes that it is ITMOs. In general, these registries have been referred to as national registries. The term of register or registry has been used. As opposed to the Kyoto Protocol, each Party is expected to have a registry. However, some Parties will argue that a registry is only necessary if the Party is purchasing not if it is a seller only.

(ii)   **An international registry for Parties that do not have or have access to such a system.** It is expected that some Parties may not have the expertise, resources, or inclination to create themselves such national registries, in which case space in

---

38   Currently, under SBSTA agenda item "methodological issues under the Paris Agreement," Parties are discussing the "Outlines of the biennial transparency report, national inventory document and technical expert review report pursuant to the modalities, procedures and guidelines for the transparency framework for action and support." [https://unfccc.int/sites/default/files/resource/sbsta2019_02E.pdf]

an international registry could be made available to function as national registry for those Parties.

(iii)   **Article 6.4 Mechanism Registry.** This is for the creation, transfer, and other actions pertaining to A6.4 expert reviews.

(iv)   **Article 6 Database.** This is where the information reported by Parties—on an annual basis in the database, and biennially through the BTR—would be kept.

(v)   **Centralized Accounting and Reporting Platform.** The Centralized Accounting and Reporting Platform (CARP) will facilitate transparency by publishing the information that Parties submit through BTRs and in the Article 6 database.

Not part of the infrastructure, but an integral part of this constellation of infrastructure and feeding information are the reports that come through the biennial update reports.

2.8.1 Article 6 National Registries

The main function of each Party's Article 6 registry is to enable Parties to record and track information, as applicable, on the transfer and acquisition of ITMOs which are:

(i)   transferred in and out,

(ii)   cancelled,

(iii)   first transferred,

(iv)   used toward the NDC, and/or

(v)   used toward other purposes.

The information in these national registries emerges bottom–up as ITMOs are authorized and transfers take place and is then captured and synthesized in the information listed above.

It must be emphasized that when discussions take place, the term "national registry" is used without a clear understanding of the functions of a registry. In general, the term registry has been used for the infrastructure that holds and tracks assets. It has been associated with serial numbers of different assets (e.g., CERs and ERUs in the Kyoto Protocol market), and transfers in relation to markets.

The discussion should clarify the difference between what a registry does in tracking assets (which is what is generally understood) and what a registry may be expected to do under the Paris Agreement—some Parties and stakeholders use the terms "register" and "registry" to distinguish these functions.

In this sense, some Parties feel that the name of registry may be misleading and a register that would keep information related to ITMOs (Section 2.1) would be more appropriate. This needs to be seen in the context that ITMOs do not have serial numbers in the classic sense under Article 6 of the Paris Agreement. Serial numbers and tracking assets will be more closely associated with registries under an ETS or different standards (Gold Standard, VERRA).

Information synthesized from registries may be stored in a UNFCCC register that is then used to report to the Article 6 database—that is, only information on those transfers that carry an Article 6.3 letter of approval will be stored in the Article 6 register.

Currently, some Parties have registries under the Kyoto Protocol. Some Parties also have or have access to domestic registries to track ETS transfers in or across their jurisdiction(s), or to track domestic or international credits. This discussion raises a number of questions that need to be clarified in future negotiating sessions:

(i)   Since these are national registries or registers, do they need to be standardized under Article 6, or each Party can develop them without any common standards, provided they enable Parties to track and deliver the needed information for the Article 6 of the Paris Agreement? The instinctive answer is "no" and the absence of any guidelines or future work will indicate that it is the correct interpretation.

(ii)   What is the relationship between registers and registries, are they one and the same, or are they different parts of the infrastructure? Can functions, including delivery of Article 6 functions under the registry, be combined in one facility with other functions, such as tracking assets? This raises additional questions.

In the case when there is one national registry (which undertakes both the function of tracking assets and the function of serving Article 6.2), this registry will track all units transferred domestically as well as those that are in and out of that country and may be called to also track the ITMO information listed in section 2.1. Capturing the information on ITMOs from the general information of transfers that the registry includes as part of its function is a procedure that needs to be determined but the information could be stored in the same registry.

In the case when there are separate pieces of infrastructure that capture information on tracking assets and another registry or register for tracking ITMOs, this implies two systems: one for tracking transfers that have Article 6.3 certification (ITMOs), and one that stores all transfers, including those with Article 6.3 certification.

 Some additional issues that should be discussed that are very relevant and also point to how discussing issues related to Article 6.2 in isolation makes the process more difficult than it needs to be.

ITMO-related information that is kept in the registry/register. How are they used by the UNFCCC Paris Agreement process? There does not seem to be any clearly enunciated use for this information in the text. Also, there is no Paris Agreement-mandated standardization.

All national decisions. Why is it then mandated by the Paris Agreement that Parties should have a national registry and track certain information, if that information is of no use to the UNFCCC process? There is no indication of direct information exchange between the registries and the Article 6 database.

A second and very relevant discussion is when a mitigation action becomes an ITMO. Although it is not clarified at this stage, this discussion needs to be elaborated to avoid a subsequent administrative decision that put Parties in front of a *fait accompli*. It is necessary to have a clear line that explains the process running from authorization, the effects on the national registry, database, etc. One scenario, but by no means the only one that can be considered, is described as follows:

(i)   When a unit or mitigation outcome receives an Article 6.3 export authorization (and therefore promise for undertaking a corresponding adjustment) and is transferred, it thereby becomes an ITMO.

(ii)   This needs to be reflected in the issuing Party's registry or register.

(iii)   It will need to be reflected in the Party's registry and the report/electronic format to the Article 6 database.

(iv)   Qualitative reporting information needs to be provided to the BTR.

(v)   Corresponding adjustment (in year x and/or target year).

(vi)   It is registered in the importing Party's registry but not its register.

(vii)   The importing Party may or may not grant it an Article 6.3 user certificate for use in its NDC. It could simply be held in the registry until such time when it is owned and used by that Party toward its NDC (and then makes the corresponding adjustment), or may simply be transferred into another registry under the same conditions.

(viii) Some Parties may allow into their registries only ITMOs that could be used toward their NDC, while other Parties may allow ITMOs in their registry that may or may not be usable toward their NDC.

(ix)   When ITMOs are used by a Party it is then recorded in its register.

2.8.2 Article 6 Database

The second important element in the infrastructure of Article 6.2, in which Parties will report the information from the Reporting section (qualitative and quantitative), is the Article 6 database.

This database can be loosely seen as accomplishing part of the functions of the International Transaction Log (ITL) under the Kyoto Protocol in that it records information on all transfers (ITMOs) but, unlike the ITL, it does not have any role to check or approve them. Other issues that need to be highlighted:

• Any tracing of ITMOs for different uses, including NDC use, will be done by using information in the database.

• There is no direct technological link or flow of information from the national registries to the database.

2.8.3 Centralized Accounting and Reporting Platform

The CARP is the overarching information platform which publishes information submitted through the Article 6 database where Parties submit annual information on transfer of ITMOs.

It makes information public about the various Article 6 infrastructure systems, including the database, the international registry, the Article 6.4 mechanism registry, and related contents from Parties' BTRs.

The Article 6 database and the CARP, in tandem, summarize the information about what is being transferred by countries and the flow of units. These transactions happen at different levels. For instance, Parties will provide information on ITMOs representing units that are transferred in the Article 6.4 Mechanism (A6.4M) registry. That same Party may also have a bilateral exchange with another Party; this is the type of summary information that is expected to be published on the CARP through the information in the database.

Although the CARP is supposed to be developed to accommodate both information required for submission under Article 6.2 guidance and information on the A6.4M Registry, further elaboration on this matter may be needed.

2.9 Ambition in Mitigation and Adaptation Actions

This section of the Article 6.2 draft text amalgamates, in what has become a generic name chapter, two issues: overall mitigation of global emissions (OMGE) and share of proceeds (SOP) for adaptation.

SOP was a provision included in the CDM that would allow a share of issued certified reduction units to be monetized for the Adaptation Fund. SOP under Article 6.2 was discussed at length at COP21 in Paris and was rejected. It was brought back without a clear "hook" in the Paris Agreement text by some Parties through political pressure.

This section has changed significantly since the text that came out of SBSTA 50 in June 2019.[39] Recalling what was in the SBSTA 50 text may be useful to readers.

SOP is one of the most contentious issues, and one of those that require political attention. It was strongly debated in Paris and not included in the Paris Agreement under Article 6.2, but has been a strong political push by some of the negotiating groups.

Positions on this issue are entrenched, with some Parties feeling strongly that it should be included, while others see it as a red line if included. Attempts at negotiating sessions to address this issue through adaptation financing, such as other sources of funding outside carbon markets, have not proved successful so far.

Therefore, the issue under debate needs to focus whether SOP under Article 6.2 is in or not, as well as on the way that it may be applied if it makes it in.

---

[39] Report of the Subsidiary Body for Scientific and Technological Advice on its 50th session, held in Bonn 17–27 June 2019. Available at https://unfccc.int/sites/default/files/resource/sbsta2019_02E.pdf.

In the text from SBSTA 50 in the options provided, there was a differentiation, with some options referring to ITMOs emanating from baseline-and-credit mechanisms, while others refer to all and any ITMO. That would be one choice. The argument can be made that imposing an SOP for ITMOs originating from baseline-and-credit mechanisms will ensure a level playing field with units produced under the A6.4M. While maybe singling out baseline-and-credit approaches and handicapping them, it is seen as a potential compromise.

A second issue was the amount of the SOP. Although 2% is included in the text, it is clearly not a final amount. Parity with the SOP under Article 6.4 is also mentioned.

The timing and the number of times that the SOP is collected was also under negotiation. Options include at first issuance, at first transfer, or every time there is a transfer, increasing with each transfer.

Some of these issues may inadvertently be resolved if the issues already mentioned are resolved (e.g., inside or outside NDC).

The current text eliminates many of these issues by encouraging Parties to

commit to contribute resources to adaptation, primarily through contributions to the Adaptation Fund, and to contribute commensurate with the rate delivered under the mechanism established by Article 6, paragraph 4, to assist developing country Parties that are particularly vulnerable to the adverse effects of climate change to meet the costs of adaptation.[40]

In fact, SOP now clearly becomes an aspirational goal.

The second issue incorporated is OMGE. The current draft text also varies significantly from the text for SBSTA 50:

Participating Parties and stakeholders are strongly encouraged to cancel ITMOs to deliver an overall mitigation in global emissions that is commensurate with the scale delivered under the mechanism established by Article 6, paragraph 4, and that is not counted towards any Party's NDC or for other international mitigation purposes.[41]

The SBSTA 50 text was a lot more elaborate. OMGE is in a very similar position as SOP, discussed above, in terms of discussions in Paris, and beyond. As in the case with SOP, the first issue in play is whether OMGE was to be included in Article 6.2. The argument which is always made is the lack of rationale to disadvantage the Article 6.4 option versus the Article 6.2 one, as OMGE is included in Article 6.4. The counter argument is always – *"we settled this one in Paris."*

The second issue at SBSTA 50 was how to operationalize OMGE in Article 6.2. The variables in this case were the basis of the overall mitigation, the timing of the OMGE, and the amount that would be considered.

---

[40] Footnote 15, Paragraph 37.
[41] Footnote 15, Paragraph 39.

This section identifies ongoing negotiation issues concerning Article 6.4 and gives a detailed account of the main points of contention.

The identified issues are related to definitions, role of the supervisory body, participation responsibilities, activity cycle (including methodologies and authorization), concept of first transfer, overall mitigation of global emissions, and CDM transition, all as contained in the version 3 of the draft text proposed by the President.

3.1 Definitions

The definitions contained in the President's proposal were streamlined from the version coming from SBSTA 50 with simple reference to Article 6 paragraphs 6.4.to 6.7 and relevant CMA decisions, and therefore quite straightforward.

The definition for the Article 6.4 reductions leaves no doubt about what is being issued: units measured in CO_2e, equal to 1 ton and in accordance with the methodologies of the IPCC adopted by the CMA.

An "Article 6, paragraph 4, emission reduction" (A6.4ER) is issued for mitigation achieved pursuant to Article 6, paragraphs 4–6, these rules, modalities and procedures, and any further relevant decisions of the CMA. It is measured in carbon dioxide equivalent and is equal to 1 tonne of carbon dioxide equivalent calculated in accordance with the methodologies and metrics assessed by the Intergovernmental Panel on Climate Change and adopted by the CMA or in other metrics adopted by the CMA pursuant to these rules, modalities and procedures.[42]

3.2 Supervisory Body

There is nothing controversial in these chapters as the governance at this level seems to be generally accepted by Parties.

---

[42] Refer to Paragraph 1 (b) of the Rules, modalities and procedures for the mechanism established by Article 6, paragraph 4, of the Paris Agreement (third iteration, 15 December), available at https://unfccc.int/documents/204686.

3.2.1 Rules of Procedure

With respect to the rules of procedure for the A6.4M, there are many provisions in the text, but the ones that have been more heavily debated and are covered by this paper are issues around Supervisory Board composition. The issues that need to be ironed out include the maximum number of terms a member can serve as a full or an alternate member, as well as the role of the alternate members in defining a quorum, and their presence and role at meetings of the supervisory body.

These issues would seem to indicate some preoccupation with the role of alternate members and the longevity of the presence of individual members on the Supervisory Board. Some Parties would seem to want to limit it to a smaller Supervisory Board. That would also reflect the options available in the draft text regarding the number of members and alternates of the Supervisory Board, with many developing countries favoring a larger Supervisory Board.

3.2.2 Governance and Functions

The supervisory body will have functions that are in line with what has been generally experienced under the CDM: establish the process required to operationalize the mechanism, support the implementation of the mechanism, and report annually to the CMA.

This will include the accreditation of operational entities, development and approval of methodologies, registration of activities, etc. These functions are also activities that the CDM Executive Board is currently undertaking such as maintaining a public website, promoting regional availability of Designated Operational Entities, providing public information to the CMA on registered activities, etc.

Some see the role of the supervisory body as currently defined to be fairly broad and in some cases outside the regulatory arena, which brings the risk of potential conflicts and also the risk that it may distract it from its regulatory duties. Functions currently proposed may include promoting public awareness, facilitating dialogue with host parties and other stakeholders, and providing annual information to the CMA on all Article 6.4 activities hosted by each Party.

3.3 Participation Responsibilities

This is one issue that has been highlighted at SBSTA 50 and 51, with some Parties making it clear that they favor a higher level of decentralization of the functions, when compared to the Executive Board of the Clean Development Mechanism (CDM EB). However, while some strongly support a more decentralized governance, others feel that the strengths of the CDM, and consequently the A6.4M, will rest with the high level of quality control of consistency in decisions that the mechanism is known for.

However, it must be emphasized that the decentralization is not envisaged as a free-for-all without any standards or supervision. The options currently put forward that would allow for such decentralization emphasize common standards and the central supervision by the Supervisory Board on any functions that may be decentralized, implying that some functions instead is performed by the Parties. This will imply a more limited role for the supervisory body.

"Functions that it intends to exercise, subject to these rules, modalities and procedures, under the supervision of the Supervisory Body, pursuant to further relevant decisions of the CMA."[43]

One issue for discussion is while the supervisory body will issue guidelines for Parties to use, it will also have the ability to overturn decisions made by a Party, if it feels that the guidelines had not been properly applied?

Outside this debate, the participation responsibilities of Parties include what can be expected: maintaining an NDC, creating a designated national authority, as well as potentially outlining baseline approaches and crediting periods to be applied to Article 6.4 activities.

Other areas of this section of the rulebook may not be as clear and their interpretation will undoubtedly become clearer only when the mechanism is operationalized, such as "Its participation in the mechanism contributes to the implementation of its NDC, and its long-term low GHG emission development strategy, if applicable."[44] Such provision could be given broad interpretation, with some seeing it as an indication that Article 6.4 activities may only take place "inside the NDC."

3.4 Article 6.4 Activity Cycle

The activity cycle is not dissimilar to that of the CDM and includes activity design, methodologies, authorization, validation, registration, monitoring, verification and certification, issuance, renewal, and voluntary cancellation.

As in previous chapters, the focus will be on those provisions that are proving controversial and need to be discussed.

3.4.1 Activity Design

The type of activities listed, including emission removals in the activities of the A6.4M, causes concerns for some Parties. While carbon capture and storage was included in the CDM after a long negotiation at the COP16 in Cancun, there are strong views that removals by sinks, as in REDD+, are under Article 5 and not under Article 6.

---

[43]   Footnote 42, Paragraph 27 (c).
[44]   Footnote 42, Paragraph 28 (b).

However, there are Parties that feel that the inclusion of REDD+ is a must, whether under Article 6.2 or Article 6.4.

Another aspect that needs to be highlighted is that a provision currently included specifies that an Article 6.4 activity "shall be designed to achieve emission reductions in the host Party."[45]

This could be easily interpreted as banning cross-border projects and cooperation as is currently the case for a number of renewable energy CDM projects, including Dagachhu Hydropower Project, Upper Marsayangdi-2 Hydro Electric Project, Félou Regional Hydropower Project, Nam Lik 1-2 Hydropower Project, and Nam Lik 1 Hydropower Project. These projects involve the export of renewable energy generated in the host country to a buying country on interconnected transnational power grids. In these cases, the activity is the generation of renewable energy in one country (host country), while the reduction takes place in another country (the importing country grid). Should this happen, export of clean energy will no longer qualify as an Article 6.4 activity.

3.4.2 Methodologies

Methodologies can be developed by a host of actors and stakeholders. Setting the baseline is a major preoccupation for many very active Parties who would like to ensure that issues such as identifying the reference year and availability of historical data are decided before they agree to set up a supervisory body which, based on the CDM Executive Board experience, may become very politicized.

A number of factors are listed as needing to be taken into account: conservative approach, consistency with the NDC of the Party, long-term goal of the Paris Agreement, long-term low-emissions development strategy of the Party, etc.

It is a long list of issues to be considered which, while certainly well-intentioned, might also result in an increased uncertainty and perceived subjectivity in setting baselines. The CDM had become complex by attaching additional features to it—such as sustainable development or environmental assessment and local participation guidance—thereby trying to solve many problems that were not always meant to be addressed by the Kyoto Protocol mechanisms. One of the reasons for its replacement with Article 6.4 was this complexity, in addition to perceived challenges related to environmental credibility from some stakeholders. The balance between environmental credibility and an operational mechanism is a fine one.

The draft text also contains provisions requiring that the supervisory body adopt principles related to additionality, as well as baseline and methodologies.

As a special provision for least-developed countries (LDCs) and small island developing states (SIDs), there is currently a provision that would waive additionality for these countries. This makes some Parties uneasy, as waiving such a critical part of the activity

---

[45]   Footnote 42, Paragraph 31 (c).

cycle, even for a good cause, may create a bad precedent. Clearly, there is a desire to make special provisions for SIDs and LDCs, but one needs to ensure that this is the right one to be included in that category.

3.4.3 Approval and Authorization

Approval and authorization is to be provided if certain conditions are met. These will include the CDM-like certification that it fosters sustainable development. The other conditions circle back to the discussion about issuing emissions reductions under Article 6.4, and the obligations that some Parties see in having a corresponding adjustment as necessary. This is a very repetitive theme that will be settled politically, under its own heading.

The host Party shall provide to the Supervisory Body the authorization for A6.4ERs issued for the activity to be internationally transferred for use towards NDCs or to be used for other international mitigation purposes or for other purposes, if the Party decides to do so, and a statement as to whether a corresponding adjustment will be applied by the host Party for A6.4ERs in accordance with chapter IX below (Avoiding the use of emission reductions by more than one Party).[46]

Another important provision is: "...the authorization of public or private entities to participate in the activity as activity participants under the mechanism."[47]

It is important to note that there is also a provision for, "other participating Parties shall provide to the Supervisory Body the authorization for public or private entities to participate in the activity as activity participants under the mechanism."[48]

This again brings in the symmetry that existed under the CDM. In those circumstances, one can argue that this was required due to the Annex 1/non-Annex 1 division, and the fact that non-Annex 1 countries did not have registries. The need and continued implications for this provision will have to be further understood.

3.4.4 Issuance

There are no special issues to signal under the issuance chapter, except how it seems to be generally accepted that the issuance will be done the same way as in the CDM, in a central mechanism registry. If that would not to be the case, and any issuance is to be done in a party registry, the argument that there is no need to make a corresponding adjustment for the first issuance under the Article 6.4 mechanism could be discussed in a different light, as the first international transfer would be between the host party and the purchasing party, and would require a corresponding adjustment.

---

[46]   Footnote 42, Paragraph 41.
[47]   Footnote 42, Paragraph 40.
[48]   Footnote 42, Paragraph 42.

The need for a central mechanism registry was never discussed and many Parties took this for granted. Under the CDM it was needed as non-Annex 1 Parties did not have registries. Under the Paris Agreement, all Parties have to have a registry so in reality, a central mechanism may not be as necessary as some may feel. Alternatively, for some Parties, having a central registry provides the excuse that the first transfer is not made to a Party and therefore a corresponding adjustment is not needed.

Another interesting provision which merits to be highlighted is:

The mechanism registry shall identify issued A6.4ERs that are authorized by the host Party for international transfer for use towards NDCs or for other international mitigation purposes or for other purposes, in accordance with the host Party's approval of the registered Article 6, paragraph 4.[49]

The question that this provision implies is if A6.4U could be issued without the Party's authorization for international use.

3.4.5 First Transfer from the Mechanism Registry

This provision brings in the 2% that the issuance will put aside for the Adaptation Fund.

It also brings in a mandatory provision for overall mitigation in global emissions: "The mechanism registry administrator shall transfer, for cancellation a percentage of the issued A6.4ERs, to the account for mandatory cancellation for delivering an overall mitigation in global emissions."[50]

It is important to note that this section of the draft text, as well as the section on "Delivering Overall Mitigation," discussed in Chapter 3.4.6, make the assumption that this is obligatory, and not voluntary or aspirational. This goes beyond the Paris Agreement which makes this aspirational as included in Article 6.4: "[...] serving as the meeting of the Parties to the Paris Agreement, and shall aim: [...] (d) To deliver an overall mitigation in global emissions."

3.4.6 Delivering Overall Mitigation in Global Emissions

This provision is put forward as obligation rather than voluntary or aspirational. It is also important to note that the overall mitigation is to be done at issuance and not at use, which is likely to lead on greater impact on sellers, who are more likely to be developing countries, at least for some time.

At issuance of A6.4ERs:

49   Footnote 42, Paragraph 52.
50   Footnote 42, Paragraph 56.

(a)   *The host Party shall make a corresponding adjustment consistent with decision X/*
        CMA.2 (guidance on cooperative approaches referred to in Article 6, paragraph 2, of the
        Paris Agreement) for the total number of issued A6.4ERs;

(b)   *The mechanism registry administrator shall transfer a percentage of the issued*
        A6.4ERs to the cancellation account in the mechanism registry for overall mitigation in
        accordance with chapter V above (Article 6, paragraph 4, activity cycle), at a level to be
        determined by the CMA that shall not be less than 2 percent....[51]

3.4.7 Avoiding the Use of Emissions Reductions by More Than One Country

This issue has been already discussed under section 1.5.5 of this paper and has been one of the main stumbling blocks in reaching an agreement on the Article 6 Paris Agreement rulebook.

While a majority of Parties feel that avoiding double counting for the first transfer of an A6.4U needs a corresponding adjustment, a significant minority of countries will argue that it is not necessary.

This requires a further attempt to explain the logic used to put forward such as position that may sound counterintuitive. The argument is based on a number of interpretations. The language used to ensure the avoidance of double counting is different between Articles 6.2 and 6.4, with no mention of corresponding adjustments under 6.4, and only reference to no use toward issuing Party NDC, if used toward another NDC.

Emission reductions resulting from the mechanism referred to in paragraph 4 of this Article shall not be used to demonstrate achievement of the host Party's nationally determined contribution if used by another Party to demonstrate achievement of its nationally determined contribution.[52]

In addition, since under Article 6.4 private sector entities will be important actors, any transfers that will take place will originate from emission reductions from outside the NDC (in addition to accomplishing the NDC) and therefore will not require a corresponding adjustment (which is done to the NDC). To do otherwise will imply that the private sector gets the benefit and the state holds the obligation toward the NDC, which then requires an increased level of effort.

All these arguments lead these Parties to conclude that an A6.4U can only be issued from outside the NDC, or in addition to meeting an NDC and therefore do not require a corresponding adjustment to avoid double counting. In this logic, an A6.4U behaves like an ITMO only starting at the second transfer (the first one is from the Article 6.4 registry to the first buying Party).

---

[51]   Footnote 42, Paragraph 67.
[52]   Footnote 1, Article 6 Paragraph 5.

It is interesting to note the attempt made by the President to bridge the gap and reach an agreement by proposing an opt-out provision for countries.

A host Party shall apply a corresponding adjustment for all A6.4ERs first transferred consistent with decision X/CMA.2 (Guidance on cooperative approaches referred to in Article 6, paragraph 2, of the Paris Agreement), subject to a future decision of the CMA that shall provide an opt out period, during which a host Party that first transfers A6.4ERs from sectors and greenhouse gases (among others) not covered by its NDC is not required to apply a corresponding adjustment.[53]

This was rejected by the proponents of the no corresponding adjustment view as putting them in the position of being seen as using what would be portrayed as an environmentally questionable clause and be subjected to pressure from civil society and other Parties to abstain. Alternatively, market pressure can be applied by refusing to buy any A6.4U from a country that uses that clause.

3.4.8 Use of Emissions Reductions for Other International Mitigation Purposes

This use is not treated in any special way than what is discussed in Chapter 3.4.7 of this paper.

3.4.9 Transition of Clean Development Mechanism Activities and Certified Emission Reductions

This has been another issue that had been very controversial and is perceived as one of the reasons why an agreement was not reached in COP24 and 25. Joint Implementation, the other project-based mechanism under the Kyoto Protocol, has not been subject to the same debate since its existence is tied to the commitment periods of the Kyoto Protocol.

It must first be mentioned that there is no reference in the Paris Agreement, or in Decision 1/CP.21 (glossary: adoption of the Paris Agreement) that justifies this topic under the Article 6 rulebook debate. The only reference to existing mechanisms is in Decision 1/CP.21, but in no way can this be invoked to justify an Article 6 discussion on this topic.

Paragraph 37 of Decision 1/CP.21 (footnote 5) states:

"Recommends that the Conference of the Parties serving as the meeting of the Parties to the Paris Agreement adopt rules, modalities and procedures for the mechanism established by Article 6, paragraph 4, of the Agreement on the basis of:
[...]

(f) Experience gained with and lessons learned from existing mechanisms and approaches adopted under the Convention and its related legal instruments [...]"

---

[53]   Footnote 42, Paragraph 70.

This is hardly a justification, but the reality is that it is a hotly debated issue with some Parties making what seems like a critical point out of it. To be fair, some Parties present this as an issue of principle in that they defend regulatory credibility and continuity, and the credibility of the UNFCCC. They argue that the same Parties have agreed to the Kyoto Protocol and the Paris Agreement, and all credibility will be lost for a regulatory market if credits, projects, and knowledge from the Kyoto Protocol mechanism are simply abandoned.

Two issues are being debated. One is what is actually the number of activities and certified emissions reductions (CERs) that are available to be transferred. Different studies show different numbers, and as always, the assumptions behind the numbers are meaningful. The numbers will vary, depending on what is counted and what is the scenario. Studies have shown that the amount of CERs that could potentially be eligible to be used post-2020 could be as high as 5 billion to 5.5 billion.[54] This number would decrease substantially if a vintage option for a projects' registration date would be introduced, e.g., the amount of CERs potentially eligible could decrease to as low as 50 million circa, if a 2016 vintage year would be applicable. Some see the numbers as being wildly exaggerated, which makes negotiations difficult.

The other one is what are the parameters or filters that can or should be used to decide which, if any of the activities or CERs are allowed in Article 6.4. The discussion seems to be guided by the numbers that are put on the table as illustrated above, with concerns expressed primarily on the impact that it would have on the prices in the carbon market. A divergent point of view is that there needs to be a regulatory, moral, or logical justification for any filter and then the numbers will be what they are. It seems that in negotiations, principles evaporated as the basis for decisions, which might explain some of the difficulties that are being experienced. They are symptoms, not causes.

The draft text includes a number of provisions:

(i)   **Transition of activities.** In this case it is clear that the approval of the host country of the CDM activity is critical, that it is needed under Article 6. This is clearly related to the need to undertake a corresponding adjustment according to the rules put in place by the CMA. Other provisions include:
   a. A deadline for transition is put forward as 31 December 2023.
   b. A6.4U can only be issued for activities post-2020 CDM.
   c. Methodologies can be applied until new ones are approved by the supervisory body.

(iv) **Certified emission reduction transition.** In this case, one of the key filters proposed is the registration date of the project, to be decided at a later time, with reductions being achieved prior to 31 December 2020. Another condition is that the CERs are used toward an NDC before 31 December 2025.

---

[54]   A. Marcu, S. Kanda, and D. Agrotti. 2020. CDM Transition: CER Availability.

For activities that are being transitioned, one important feature that needs to be highlighted is that host countries do not require a corresponding adjustment, while the using Parties will have to undertake a corresponding adjustment. While not totally clear, this provision may seem to be somewhat not aligned, in principle, with the provision that ensures the avoidance of use by more than one Party, given the host parties for A6.4U will have to apply corresponding adjustment as per the decision of the CMA, with an opt-out allowed. A6.4U and CERs seem to be treated differently, even if both fly under the A6.4M flag.

4. Article 6.8: Negotiation Issues

The framework for nonmarket approaches is seen as absolutely necessary by some Parties, but it is the one element of Article 6 that is less mature than the others. It has, however, evolved since the Katowice COP and new elements have emerged that give some additional precision. All Parties seem to agree that having a positive outcome for Article 6.8 is critical if we are to have an overall positive outcome for the Article 6 rulebook in Madrid.

Article 6.8 was initially introduced to counterbalance what was perceived as the pressure from some Parties to include market approaches. However, Parties have found it challenging to translate this into a more concrete reality. That is why Article 6.8 is currently focused on two issues: defining the governance of the framework for non-markets and outlining a work program for the coming years to accompany the framework. By and large, the situation can be summarized as there is no agreement on the former and a lack of concreteness on the later.

Article 6.8 starts by elaborating principles on the framework and elements of the work program. The language would indicate that Parties are struggling with the definition of non-markets. That is not necessarily surprising since most of the elements and approaches under the UNFCCC can be deemed to be non-market. However, what best outlines the function of the framework is the following provisions in paragraphs 1(a) (i) and (ii) in the President's proposal:[55]

(i)   Facilitates the use and coordination of NMAs in the implementation of Parties' NDCs in the context of sustainable development and poverty eradication;

(ii)   Enhances linkages and creates synergies between, inter alia, mitigation, adaptation, finance, technology development and transfer, and capacity building, while avoiding duplication with the work of the subsidiary and constituted bodies under the Convention, the Kyoto Protocol and the Paris Agreement.

One important element which is a tracer for the thinking of the proponents of Articles 6.8 to 6.9 is best illustrated by the provision that "voluntary collective actions that are not reliant on market-based approaches and that do not include transactions or quid pro quo operations."[56] The preoccupation to show that cooperation under the Paris Agreement can be, and is mostly, through NMAs, is important.

---

[55]   Refer to Paragraph 1 (a) of the Work program under the framework for NMAs referred to in Article 6, paragraph 8, of the Paris Agreement (third iteration, 15 December), available at https://unfccc.int/documents/204667.

[56]   Footnote 55, Paragraph 1 (b).

4.1 Nonmarket Approaches under the Framework

The type of activities listed in the President's proposal could indicate a very broad coverage at this point and as illustrated by the following points: "(i) Mitigation, adaptation, finance, technology development and transfer, and capacity- building, as appropriate; (ii) Contribution to sustainable development and poverty eradication in participating Parties".[57]

4.2 Governance of the Framework

The issue that is discussed is that of the governance of the framework with two main options. One is outlined in paragraph 6 where the decision to establish a permanent body to manage the framework is pushed in the future:

The subsidiary bodies will consider whether institutional arrangements for the framework that will supersede the NMA forum are needed and make recommendations for consideration and adoption by the Conference of the Parties serving as the meeting of the Parties to the Paris Agreement (CMA) at CMA 8 (2025).[58]

While some Parties see the importance and urgency of establishing a body to coordinate the NMAs, many Parties resist the establishment of new permanent bodies or structures in general, as was the case also under the topic of response measures where the establishment of the Katowice Committee of Experts on the Impacts of the Implementation of Response Measures.

The argument is also that it is difficult to make the case for establishing new structures until it is clear how the work program of the framework is defined.

4.3 Work Program of the Framework

The proposed modalities of the work program include items that are not new, such as submissions and technical papers.

The work program as presented in the draft includes a fairly general list of activities that also points to the divergence of views and some uncertainty from many Parties as to what Articles 6.8 to 6.9 are expected to have as output, considering that many of the NMAs under the UNFCCC and CMA are already up and running, some with program. Also, another way to put this is—what does Article 6.8 bring to any NMA that would make it

---

[57]   Footnote 55, Paragraphs 2 (b) (i) and (ii).
[58]   Footnote 55, Paragraph 6.

compelling for someone to use its provisions? The two main components of the work program are

(i)   identifying measures to enhance existing linkages, create synergies and facilitate coordination and implementation of NMAs; and

(ii)   implementing measures.

(i) Developing and implementing tools, including a UNFCCC web-based platform for recording and exchanging information on NMAs

(ii) Identifying and sharing relevant information, best practices, lessons learned and case studies for developing and implementing NMAs, including on how to:

(a)   *Replicate successful NMAs, including in the local, subnational, national and global context;*

(b)   *Facilitate enabling environments and successful policy frameworks and regulatory approaches;*

(c)   *Enhance the engagement in NMAs of the private sector, and vulnerable and impacted sectors and communities;*

(d)   *Leverage and generate mitigation co-benefits that assist the implementation of NDCs.*[59]

One of the fundamental questions is to what degree the provisions in these articles are action-oriented, or are more facilitative in nature. So far, the draft would indicate a more facilitative and coordination role, including in some way ensuring that the current NMAs are coordinated and optimized.

Some Parties and stakeholders have made efforts to ensure that there are substantive offers on the table, but have come with controversial proposals. One proposal put forward and which originated with the African Development Bank proposes an adaptation mechanism, which looks into contributions to adaptation and provides adaptation credits.

Another approach would see Article 6.8 play a role in ensuring that the socioeconomic impacts that emerge through Articles 6.2 and 6.4 would be addressed by giving a mandate to the NMA Forum to liaise with the Forum on Response Measures and act to address them.

A last proposal put forward was that Article 6.8 would be used as the basis for ensuring that the financial contribution of developed countries to developing countries would be recognized and measured, but without any credits flowing back to the donor country. It would be a measure to recognize the additional contribution, beyond their own NDC, that developed countries have made toward global mitigation.

In the final analysis it is accepted by all Parties that Article 6.8 is an essential part of the Article 6 rulebook package and development will need to be shown if there is to be progress on the overall Article 6 rulebook.

---

[59]   Footnote 55, Paragraphs 8 (b) (i) and (ii).

5. Relation with Other Issues

5.1 Brief Introduction to Monitoring, Reporting, and Verification under the Paris Agreement

Article 13 of the Paris Agreement has established an enhanced transparency framework (ETF) for monitoring, reporting and verification of actions and support related to climate change (Figure 2).

Figure 2: Article 13 of the Paris Agreement—Transparency of Action and Support

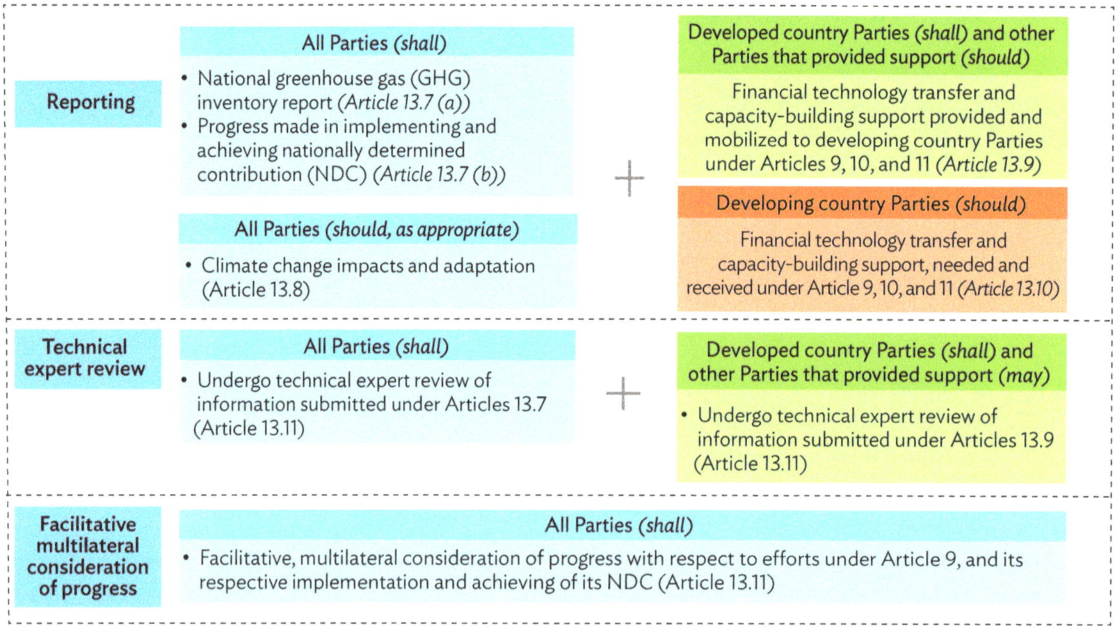

Reporting

All Parties (shall)
- National greenhouse gas (GHG) inventory report (Article 13.7 (a))
- Progress made in implementing and achieving nationally determined contribution (NDC) (Article 13.7 (b))

All Parties (should, as appropriate)
- Climate change impacts and adaptation (Article 13.8)

+

Developed country Parties (shall) and other Parties that provided support (should)
Financial technology transfer and capacity-building support provided and mobilized to developing country Parties under Articles 9, 10, and 11 (Article 13.9)

Developing country Parties (should)
Financial technology transfer and capacity-building support, needed and received under Article 9, 10, and 11 (Article 13.10)

Technical expert review

All Parties (shall)
- Undergo technical expert review of information submitted under Articles 13.7 (Article 13.11)

+

Developed country Parties (shall) and other Parties that provided support (may)
- Undergo technical expert review of information submitted under Articles 13.9 (Article 13.11)

Facilitative multilateral consideration of progress

All Parties (shall)
- Facilitative, multilateral consideration of progress with respect to efforts under Article 9, and its respective implementation and achieving of its NDC (Article 13.11)

- The transparency framework shall provide felixibility in the implementation of the provisions of the Article to those developing country Parties that need it in the light of their capacities (Article 13.2)
- The transparency framework shall recognize the special circumstances of the least-developed countries and small island developing States (Article 13.3)

Source: UNFCCC. 2020. Reporting and Review under the Paris Agreement. https://unfccc.int/process-and-meetings/transparency-and-reporting/reporting-and-review-under-the-paris-agreement.

Under the ETF, all Parties of the Paris Agreement will have to submit BTRs[60] with, among others, information necessary to track progress made in implementing and achieving its NDC that includes

(i)   description of its national circumstances relevant to progress made in implementing and achieving its NDC;
(ii)   description of its NDC;
(iii) indicators (qualitative or quantitative) that have been selected to track progress toward the implementation and achievement of its NDC;
(iv)   information on actions, policies, and measures that support the implementation and achievement of its NDC;
(v)   summary of GHG emissions and removals; and
(vi)   projections of GHG emissions and removals, as applicable.

Such information shall be presented according to the reporting requirements established by Decision 18/CMA.1: MPG for the transparency framework for action and support referred to in Article 13 of the Paris Agreement.

In particular, in relation to the indicators, Parties shall provide the information in a "structured summary," that includes, for each selected indicator:

(i)   Information for the reference point(s), level(s), baseline(s), base year(s), or starting point(s);
(ii)   Information for previous reporting years during the implementation period of its NDC, as applicable; and
(iii) The most recent information.[61]

When a Party participates in cooperative approaches that involve the use of ITMOs toward an NDC, or authorizes the use of mitigation outcomes for international mitigation purposes other than the achievement of its NDC, Decision 18/CMA.1 requires that it shall also provide the following information in the structured summary:

consistently with relevant decisions adopted by the CMA on Article 6:

(i)   The annual level of anthropogenic emissions by sources and removals by sinks covered by the NDC on an annual basis reported biennially;
(ii)   An emissions balance reflecting the level of anthropogenic emissions by sources and removals by sinks covered by its NDC adjusted on the basis of corresponding adjustments undertaken by effecting an addition for internationally transferred mitigation outcomes first-transferred/transferred and a subtraction for internationally transferred mitigation outcomes used/acquired, consistent with decisions adopted by the CMA on Article 6;
(iii) Any other information consistent with decisions adopted by the CMA on reporting under Article 6; and

---

[60]   The first BTR is to be submitted at the latest by 31 December 2024 (paragraph 3 of Decision 18/CMA.1).
[61]   Footnote 9, Paragraph 77 (a) (i–iii).

(iv)   Information on how each cooperative approach promotes sustainable development; and ensures environmental integrity and transparency, including in governance; and applies robust accounting to ensure inter alia the avoidance of double counting, consistent with decisions adopted by the CMA on Article 6.[62]

Information provided, in the BTR to track progress made in implementing and achieving its NDC will be subjected to a TER, according to rules defined by Decision 18/CMA.1.

Currently, Parties are negotiating under the SBSTA agenda "methodological issues under the Paris Agreement" the "common tabular formats" for the electronic reporting of the information necessary to track progress made in implementing and achieving the NDC, which includes the "structured summary." That means that under SBSTA "methodological issues under the Paris Agreement", the "format" of the "structured summary" is under discussion; not necessarily its content. SBSTA is expected to conclude its consideration on these matters by COP26 and forward draft text decisions for adoption by CMA 3 (Glasgow).[63]

5.2 Links between Article 6 and the Enhanced Transparency Framework

As can be noted, besides the fact that the ETF MPG (Decision 18/CMA.1) have already decided what information need to be included in the structured summary, details of some of the information to be presented are linked with decisions adopted by the CMA on Article 6, in particular:

(i)   guidance for corresponding adjustments;
(ii)   information on how each cooperative approach promotes sustainable development and ensures environmental integrity and transparency, including in governance, and applies robust accounting to ensure, among others, the avoidance of double counting; and
(iii) any other information required by decisions adopted by the CMA on Article 6.

5.2.1 Corresponding Adjustments Guidance

For Parties that will not participate in cooperative approaches, it can be argued that all information to be presented in the structured summary, to demonstrate progress made in implementing and achieving its NDC, have already been listed in Decision 18/CMA.1.

In these cases, the information will be mainly related to the indicators selected by the Party and the accounting approach:

---

[62]   Footnote 9, Paragraph 77 (d) (i-iv).
[63]   The latest outcomes of these negotiations can be found in the Report of the SBSTA on its 50th session, held in Bonn from 17–27 June 2019, paragraphs 115–129.

(i)   For the first NDC, each Party shall clearly indicate and report its accounting approach, including how it is consistent with Article 4, paragraphs 13 and 14 of the Paris Agreement. Each Party may choose to provide information on accounting of its first NDC consistent with decision 4/CMA.1.[64]

(ii)   For the second and subsequent NDC, each Party shall provide information consistent with decision 4/CMA.1. Each Party shall clearly indicate how its reporting is consistent with decision 4/CMA.1.[65]

Decision 4/CMA.1[66] has defined, through its Annex II, the rules for NDC accounting. None of the rules are specifically targeted to Article 6 and/or ITMOs.

Regardless of the accounting approach chosen by the Party, in each BTR that contains information on the end year of the NDC implementation period, the structured summary needs to include an assessment of whether it has achieved the target(s) for its NDC (para 23(h) of the Article 6.2 text).

Only in the cases where a Party has decided to participate in cooperative approaches that involves the use of ITMOs, the information on the structure summary, including the assessment of whether it has achieved the target(s) for its NDC, will need to be presented consistently with the guidance defined under Article 6 CMA decisions, in particular guidance related to corresponding adjustments.

For those cases, it is worth to highlight that under the ETF MPG, an emissions balance of the GHG emissions and removals covered by the NDC is to be presented; and in this balance, the basic guidance for the "corresponding adjustment" have already been defined: "[...] undertaken by effecting an addition for internationally transferred mitigation outcomes first-transferred/transferred and a subtraction for internationally transferred mitigation outcomes used/acquired."[67]

Pending on how guidance on corresponding adjustments will be decided under Article 6, in particular in relation to the treatment of non-GHG metric, single-year NDC/multiyear NDC, sectors and GHG not covered by the NDC, other international mitigation purposes; and limits to the transfer and use of ITMOs, some changes and/or additional information will have to be added in the ETF structure summary in relation to the emissions balance. In this sense, the final format of the structure summary for Parties participating in cooperative approaches (to be decided in Glasgow) is dependent on decisions to be taken on Article 6 (also in Glasgow).

---

[64]   Footnote 9, Paragraph 71.
[65]   Footnote 9, Paragraph 72.
[66]   Footnote 7, Decision 4/CMA.1.
[67]   Footnote 9, Paragraph 77 (d) (ii).

5.2.2 Reporting and Review under Article 6

In addition to the information related to corresponding adjustments, the ETF MPG requires Parties to report, in its structure summary in the BTR, how each cooperative approach promotes sustainable development; and ensures environmental integrity and transparency, including in governance; and applies robust accounting to ensure inter alia the avoidance of double counting. No further guidance is given in the ETF MPG on the level of detail and/or specificity of such information. This has been discussed in more detail in Chapter2.7.

6. Road to the 26th Session of the Conference of the Parties

A lot has happened since Volume I of the *Decoding Article 6 of the Paris Agreement* was published, not least of all COP24 in Katowice and COP25 in Madrid. The current hiatus in negotiations induced by measures to address the medical situation have added to the problems (e.g., there are no negotiating session, no other opportunities for face-to-face informal discussions, etc.), or maybe is allowing much needed time to mature the discussion.

Parties have gone through many iterations of proposed text for the Article 6 rulebook. By examining these texts, one may conclude that the overall content and format has not undergone enormous changes. There have been subtle changes, but from the evolution of the text one may conclude that the topic is maturing. In the final analysis, the Paris Agreement has had the great merit of bringing all Parties to the table by providing the flexibility of NDCs.

However, from a market structure point of view, especially accounting, the Kyoto Protocol was a much easier document to work with. If the diversity of the NDC is the strength of the overall Paris Agreement, it is also a weakness in bringing together such different approaches in a market. It is therefore not surprising that the negotiations are taking longer than was initially expected.

One big step forward has been the increased consensus on what the difficult issues are. To address these issues, one must first agree on what they are. A second significant step, and this has emerged largely only after SBSTA 50, is the increased acceptance that there are a number of issues, which one may call issues of principle (or with political implications) that are unlikely to be addressed at the negotiator level and will require intervention by the head of delegation, or maybe even ministerial attention.

While all issues are important, some of the more difficult ones include:

(i) What is an ITMO? What are the attributes or characteristics of an ITMO? This will include issues such as:
   a. form,
   b. metric,
   c. inclusion of sinks, and
   d. Article 6.4 mechanism (A6.4M) units.

(ii) Corresponding adjustment:
   a. What gets adjusted: NDC or emissions-based number? Or both—one with an eye on global goal, one with a Party goal?

        b. Timing of adjustment: at first transfer, at usage

        c. Amount of adjustment: treatment of single-year NDC

(iii) How to include use of ITMOs for purposes other than NDCs?

(iv) How to decide on governance and rules of procedures for Article 6.4, including level of decentralization?

(v) What information is made available to the international regulator (CMA)? What gets reported and recorded, timing of the reporting?

(vi) Is there a share of proceeds for Article 6.2?

(vii) Nature of overall mitigation of global emissions (OMGE) for Article 6.4—voluntary or not?

(viii) Is there an OMGE provision for Article 6.2 at all?

(ix) Avoidance of double counting for Article 6.4.

(x) Transition of the Kyoto Protocol mechanisms to Article 6 of the Paris Agreement.

What needs to be identified are those issues where (i) the lack of outcome is on the critical path and blocks decisions on a "bucket" of operational issues; and (ii) different outcomes may lead, in the view of some Parties, to implications for the fundamentals of the Paris Agreement.

The least promising pathway to COP26 is one that assumes that some issues have been agreed in COP24 and COP25 and that there is a very limited number of issues to discuss, with some taking the position that it would be counter productive to discuss other issues as it would reopen issues that are "settled." This ignores the "package" approach that Parties take to negotiations and there are issues that were not brought up in the discussion for the last versions of the Presidency text in COP25 as Parties agreed that it was not going anywhere and decided that it was not worth raising them.

Article 6 of the Paris Agreement recognizes that some Parties choose to pursue voluntary cooperation in the implementation of their NDCs to allow for higher ambition in their mitigation and adaptation actions, and to promote sustainable development and environmental integrity.

Assigned amount unit (AAU). A Kyoto Protocol unit equal to 1 metric ton of carbon dioxide (CO_2) equivalent. AAUs are issued for Annex 1 Parties up to the level of its assigned amount, established pursuant to Article 3, paragraphs 7 and 8 of the Kyoto Protocol. AAUs may be exchanged through emissions trading.

Bracketing using typographical symbols of square brackets [--] placed around text under negotiation to indicate that the language enclosed is being discussed but has not yet been agreed upon.

Conference of the Parties (COP). The supreme body of the Convention. It currently meets once a year to review the Convention's progress. The word "conference" is not used here in the sense of "meeting," but rather of "association". The Conference meets in sessional periods, for example, the "fourth session of the Conference of the Parties."

Conference of the Parties serving as the meeting of the Parties to the Paris Agreement (CMA). All states that are Parties to the Paris Agreement are represented at the CMA, while states that are not Parties participate as observer Parties. The CMA oversees the implementation of the Paris Agreement and takes decisions to promote its effective implementation.

Conference of the Parties serving as the Meeting of the Parties to the Kyoto Protocol (CMP). The Convention's supreme body is the Conference of the Parties (COP), which serves as the meeting of the Parties to the Kyoto Protocol. The sessions of the COP and the CMP are held during the same period to reduce costs and improve coordination between the Convention and the Protocol.

Contact group. An open-ended meeting that may be established by the COP, a subsidiary body or a Committee of the Whole wherein Parties may negotiate before forwarding agreed text to a plenary for formal adoption. Observers and observer Parties generally may attend contact group sessions.

Decision 1/CP.21 mandated the Subsidiary Body for Scientific and Technological Advice (SBSTA) to operationalize the provisions of Article 6 of the Paris Agreement through recommending a set of decisions to the COP.

Drafting group. A smaller group established by the president or a chair of a Convention body to meet separately and in private to prepare draft text, which must still be formally approved later in a plenary session. Observers generally may not attend drafting group meetings.

Informal contact group. A group of delegates instructed by the president or a chair to meet in private to discuss a specific matter in an effort to consolidate different views, reach a compromise, and produce an agreed proposal, often in the form of a written text.

Internationally Transferred Mitigation Outcomes (ITMOs). Mitigation outcomes in one Party transferred to another Party. They may or may not be used by the receiving Party toward its NDC.

Kyoto mechanisms. Three procedures established under the Kyoto Protocol to increase the flexibility and reduce the costs of making greenhouse gas emissions cuts: Clean Development Mechanism, emissions trading, and joint implementation. They are the framework for the creation of a carbon market.

Nationally Determined Contribution (NDC) to the Paris Agreement. According to Article 4 paragraph 2 of the Paris Agreement, each Party shall prepare, communicate and maintain successive NDCs that it intends to achieve. Parties shall pursue domestic mitigation measures, with the aim of achieving the objectives of such contributions.

Paris Agreement rulebook/Katowice Climate Package/COP24 decisions. Set of implementing decisions of the 2015 Paris Agreement concluded during COP24 in Katowice, Poland on 2–14 December 2018. The implementation guidelines covered informing NDCs, communication on adaptation, rules of functioning of the Transparency Framework, facilitation and compliance, global stocktake, technology transfer, and financial support.

Regional groups. Groups of Parties, in most cases sharing the same geographic region, which meet privately to discuss issues, sometimes form joint positions, and nominate bureau members and other officials for activities under the Convention. The five regional groups are Africa, Asia, Central and Eastern Europe (CEE), Latin America and the Caribbean (GRULAC), and the Western Europe and Others Group (WEOG).

Registries, registry systems. Electronic databases that tracks and records all transfers under the Kyoto under mechanisms such as the Clean Development Mechanism, joint implementation and emissions trading . Registries may have a different function under the Paris Agreement, yet to be defined. Under the Paris Agreement, they are generally seen as part of the infrastructure needed for tracking and the avoidance of double counting.

Sink. Any process, activity, or mechanism that removes a greenhouse gas, an aerosol, or a precursor of a greenhouse gas from the atmosphere. Forests and other vegetation are considered sinks because they remove carbon dioxide through photosynthesis.

Subsidiary body. A United Nations Framework Convention on Climate Change (UNFCCC) body that assists and prepares decision of the COP. Two permanent subsidiary bodies are created by the Convention: the Subsidiary Body for Implementation (SBI) and the Subsidiary Body for Scientific and Technological Advice (SBSTA).

Subsidiary Body for Implementation (SBI). The SBI makes recommendations on policy and implementation issues to the COP and, if requested, to other bodies.

Subsidiary Body for Scientific and Technological Advice (SBSTA). The SBSTA serves as a link between information and assessments provided by expert sources (such as the Intergovernmental Panel on Climate Change) and the COP, which focuses on setting policy.[68]

68   UNFCCC. Glossary of Climate Change Acronyms and Terms. https://unfccc.int/process-and-meetings/the-convention/glossary-of-climate-change-acronyms-and-terms#s.

Clean Development Mechanism Policy Dialogue. 2012. *Climate Change, Carbon Markets and the CDM: A Call to Action Report of the High-Level Panel on the CDM Policy Dialogue.* Bangkok: UNFCCC.

Initiative for Climate Action Transparency. 2020. *Transformational Change Methodology: Assessing the Transformational Impacts of Policies and Actions.* Copenhagen: Initiative for Climate Action Transparency.

Intergovernmental Panel on Climate Change (IPCC). 2018. *Global Warming of 1.5°C: An IPCC Special Report on the Impacts of Global Warming of 1.5°C above Pre-Industrial Levels and Related Global Greenhouse Gas Emission Pathways, in the Context of Strengthening the Global Response to the Threat of Climate Change.* In Press.

Marcu, A., S. Kanda, and D. Agrotti. 2020. *CDM Transition: CER Availability.*

Olsen, K. H. 2007. *The Clean Development Mechanism's Contribution to Sustainable Development: A Review of the Literature.* Climatic Change. 84 (1). pp. 59–73.

Sutter, C. and J. C. Parreño. 2007. *Does the Current Clean Development Mechanism (CDM) Deliver its Sustainable Development Aim? An Analysis of Officially Registered CDM Projects.* Climatic Change. 84 (1). pp. 75–90.

United Nations. 1995. *Kyoto Protocol to the United Nations Framework Convention on Climate Change.* Kyoto.

United Nations. 2015. *The Paris Agreement.* Paris.

United Nations Framework Convention on Climate Change (UNFCCC). 2016. *Report of the Conference of the Parties on its Twenty-first Session, held in Paris from 30 November to 13 December 2015: Part Two: Action Taken by the Conference of the Parties at its Twenty-first Session. 29 January.* Paris

UNFCCC. 2018. *Process and Meetings. Katowice Climate Package.* Katowice.

UNFCCC. 2019. *Annex to Matters Relating to Article 6 of the Paris Agreement: Rules, Modalities and Procedures for the Mechanism Established by Article 6, Paragraph 4, of the Paris Agreement. Proposal by the President. Third Iteration.* Madrid

UNFCCC. 2019. *Draft Text: Annex to Matters Relating to Article 6 of the Paris Agreement: Guidance on Cooperative Approaches Referred to in Article 6, Paragraph 2, of the Paris Agreement. Proposal by the President. Third Iteration.* Madrid

UNFCCC. 2019. *Matters relating to Article 6 of the Paris Agreement, Proposal by the President, Draft decision -/CMA.2.* Madrid

UNFCCC. 2019. *Matters Relating to Article 6 of the Paris Agreement: Work Programme under the Framework for Non-market Approaches Referred to in Article 6, Paragraph 8, of the Paris Agreement. Third Iteration.* Madrid

UNFCCC. 2019. *Modalities, Procedures and Guidelines for the Transparency Framework for Action and Support Referred to in Article 13 of the Paris Agreement.*

UNFCCC. 2019. *Report of the Conference of the Parties Serving as the Meeting of the Parties to the Paris Agreement on the Third Part of its First Session, held in Katowice from 2 to 15 December 2018.* Katowice

UNFCCC. 2019. *Report of the Conference of the Parties serving as the meeting of the Parties to the Paris Agreement on its second session, held in Madrid from 2 to 15 December 2019.* Madrid

UNFCCC. 2019. *Report of the Subsidiary Body for Scientific and Technological Advice on its Fiftieth Session, held in Bonn from 17 to 27 June 2019.*